HEALTHCARE SYSTEMS DESIGN OF INTELLIGENT TESTING CENTERS

HEALTHCARE SYSTEMS DESIGN OF INTELLIGENT TESTING CENTERS

Latest Technologies to Battle Pandemics such as COVID-19

TAWANDA MUSHIRI
Senior Lecturer, Department of Biomedical Informatics and Biomedical Engineering, Faculty of Medicine and Health Sciences, University of Zimbabwe, Mt Pleasant, Harare, Zimbabwe
Senior Research Associate, Biomedical Engineering and Healthcare Technology (BEAHT) Research Centre, Faculty of Health Sciences, University of Johannesburg, Johannesburg, South Africa

MARVELLOUS MOYO
Department of Mechatronics and Industrial Engineering, Faculty of Engineering and the Built Environment, University of Zimbabwe, Harare, Zimbabwe

ELSEVIER

ACADEMIC PRESS
An imprint of Elsevier

Academic Press is an imprint of Elsevier
125 London Wall, London EC2Y 5AS, United Kingdom
525 B Street, Suite 1650, San Diego, CA 92101, United States
50 Hampshire Street, 5th Floor, Cambridge, MA 02139, United States
The Boulevard, Langford Lane, Kidlington, Oxford OX5 1GB, United Kingdom

ISBN: 978-0-323-99443-9

For information on all Academic Press publications
visit our website at https://www.elsevier.com/books-and-journals

Publisher: Mara E. Conner
Acquisitions Editor: Sonnini R. Yura
Editorial Project Manager: Aera F. Gariguez
Production Project Manager: Fizza Fathima
Cover Designer: Matthew Limbert

Typeset by STRAIVE, India

Contents

CHAPTER 1

Introduction to artificial intelligence and robotics in healthcare systems

1.1 Introduction

Artificial intelligence was first introduced by John McCarthy in 1956 (Singh, Mishra, & Sagar, 2013) and it is the science of building computer programs that perceive their environment and take actions which would require intelligence if they were done by human beings (Cross et al., 2009). AI includes robotics, machine learning, and fuzzy logic with robotics being the main branch. Fuzzy logic involves mathematical approach to problem solving (Singh et al., 2013), and it produces exact results from imprecise data and is mostly used in computers and electronic applications. In contrast to classical logic, in fuzzy logic statements are no longer true or false, on or off, but an object can assume any real value between 0 and 1. Machine Learning (ML) is the study of making a computer act without coding. Machine Learning (ML) consists of two steps: first step involves the learning phase that uses the input data to find the parameters that best fit the task at hand. The second step will use the learned parameters as input and performs the required task. Therefore, from the foregoing statements, it can be seen that incorporating artificial intelligence and robotics will help overcome challenges caused by novel coronavirus quickly (Fig. 1.1).

1.2 Background

Robotics was first introduced in the late 1940s and since then human and robot integration has been a popular research interest (Kapila, 2010; Terris et al., 2014). Not, at this point restricted to the modern automation (assembly and manufacturing) (Brogardh, 2007; Dung et al., 2007), robotic arms can now be used in space research (Blacker & Gao, 2018; Jayaweera & Webb, 2010; Wilcox, Ambrose, & Kumar, 2008), service industry (Kim & Song, 2014), and medical field (DiMaio, Hanuschik, & Kreaden, 2011).

Healthcare Systems Design of Intelligent Testing Centers
https://doi.org/10.1016/B978-0-323-99443-9.00009-7

Fig. 1.1 A depiction of artificial intelligence and robotics. *(Data from Kindel Media. (2021). Full shot of robot toy. Studies in health technology and informatics. https://doi.org/10.3233/978-1-60750-706-2-716.)*

According to the Robotics Institute of America, a robot is a reprogrammable (Ben-Ari & Mondada, 2018) multifunctional arm intended to move devices, materials, parts, or unique gadgets through factor customized movements for the exhibition of an assortment of tasks (Kumar, 1997). Therefore the proposed definition in the medical field context is: An electromechanical tool having one or more axes which are actuated semiautonomously under the guidance of a physician to treat patients (Hanumara, 2012).

The application of robotics in medical field ranges from nanobots used to attack pathogens in the blood stream to autonomous surgical interventions (Azar, 2012; Ben-Ari & Mondada, 2018). Interventional medicine is still predominantly manual, that is, it relies more on operator's skill and dexterity, but where physicians encounter limitations, opportunities for adapting robots can be identified (Hanumara, 2012).

Table 1.1 Strengths and limitations of robots and humans (Soleimani et al., 2011).

Human strengths	Human limitations	Robot strengths	Robot limitations
Strong hand-eye coordination	Limited dexterity outside natural scale	Good geometric accuracy	No judgment
Dexterous	Prone to tremor and fatigue	Stable and untiring	Unable to use qualitative information
Flexible and adaptable	Limited geometry accuracy	Scale motion	Expensive
Can integrate extensive and diverse information	Limited ability to use qualitative information	Can use diverse sensors in control	Technology in flux
Rudimentary haptic abilities	Limited sterility	May be sterilized	More studies needed
Able to use qualitative information	Susceptible to radiation and infection	Resistant to radiation and infection	
Good judgment			

Table 1.1 summarizes the strengths and limitations of robots and humans (Soleimani et al., 2011). Medical robots are not only used to address the earlier mentioned limitations but also to help keep up with the changing world as new threats to human lives continue to increase such as the novel coronavirus (Murphy, 2020).

1.2.1 The novel coronavirus (COVID-19)

In December 2019, a novel virus (COVID-19) was first secluded from three patients with pneumonia, associated with the bunch of intense respiratory disease cases from Wuhan, China (WHO, 2020b). As indicated by the hereditary examination, this infection is firmly identified with SARS (Severe Acute Respiratory Syndrome) (World Health Organization, 2020d; Zhu et al., 2020). The principal study (published 24 January 2020) showed that the vast majority of the 41 hospitalized patients had myalgia, cough, fever, or exhaustion (Huang et al., 2020). Likewise, reports show that there is person-to-person transmission of the infection through the liquid radiated from a contaminated individual (Wang et al., 2020) and as of late a few investigations have shown that the infection is additionally airborne (Shen & Li, 2020). Also, the virus can survive for a few days away from the host. COVID-19 virus is characterized as a group of infections that range from the normal cold to MERS (Middle East Respiratory Syndrome) and SARS (Severe Acute Respiratory Syndrome) (Letko & Munster, 2020; WHO, 2020a; World Health Organization, 2020d).

1.2.2 Testing COVID-19 virus

To moderate the spread of coronavirus requires finding and testing all presumed cases so that affirmed cases are quickly and successfully confined and get proper consideration (Welch, 1995; World Health Organization, 2020a, 2020c). This will empower the nearby contacts of all affirmed cases to be distinguished quickly with the goal that they can be isolated and medicinally checked for the 14-day brooding time of the infection.

To oversee and control the outbreak, fast assortment and testing of suitable specimens from patients (see Tables 1.2 and 1.3) meeting the presumed case definition for COVID-19 is fundamental and ought to be directed by a research facility master.

Medical services workers who gather samples ought to hold fast thoroughly to contamination avoidance and control rules (World Health Organisation, 2020; World Health Organization, 2020b). Extra clinical samples might be gathered as COVID-19 infection has been identified in blood and stool, as had the coronaviruses liable for SARS and MERS (Xu et al., 2020; Zhang et al., 2020).

1.3 Motivation

1.3.1 Challenges at Sally Mugabe Central Hospital due to COVID-19 pandemic

Since the beginning of the first lockdown which started in March 2020 in Zimbabwe, Sally Mugabe Central Hospital, which is the country's biggest hospital, has been facing a lot of challenges. Some of the challenges include shortages of protective personal equipment, electricity, and basic medicine (Mahomed, 2021). Additionally, because of the absence of sufficient hardware and working space at Sally Mugabe Central Hospital (also at Mpilo Central Hospital), there has been a slow pace in the test for COVID-19 samples (Kubatana.net, 2020). For example, the National Tuberculosis Reference Laboratory (NTRL) has been specialized for testing TB and now also serves as a COVID-19 testing facility, compelling groups to work in shifts

Table 1.2 Samples to be retrieved from symptomatic patients and contacts (Welch, 1995; World Health Organization, 2020c).

	Test	Type of sample	Timing
Patient	NAAT	Lower respiratory tract - sputum - aspirate - lavage Upper respiratory tract - nasopharyngeal and - oropharyngeal swabs - nasopharyngeal wash/ nasopharyngeal aspirate. Consider stools, whole blood, urine, and if diseased, material from autopsy	Collect on presentation. Possibly repeated sampling to monitor clearance. Further research needed to determine effectiveness and reliability of repeated sampling
Patient	Serology	Serum for serological testing once validated and available	Paired samples are necessary for confirmation with the initial sample collected in the first week of illness and the second ideally collected 2–4 weeks later (optimal timing for convalescent sample needs to be established)
Contact in healthcare center-associated outbreaks or other settings where contacts have symptoms or where asymptomatic contacts have had high-intensity contact with a COVID-19 case	NAAT	Nasopharyngeal and oropharyngeal swabs	Within incubation period of last documented contact
	Serology	Serum for serological testing once validated and available	Baseline serum taken as early as possible within incubation period of contact and convalescent serum taken 2–4 weeks after last contact (optimal timing for convalescent sample needs to be established)

Table 1.3 Specimen collection and storage (Druce et al., 2011; World Health Organisation, 2018).

Specimen type	Collection materials	Storage temperature until testing in-country laboratory	Recommended temperature for shipment according to expected shipment time
Nasopharyngeal and oropharyngeal swab	Dacron or polyester flocked swabs[a]	2–8°C	2–8°C if ≤5 days −70°C (dry ice) if >5 days
Bronchoalveolar lavage	Sterile container[a]	2–8°C	2–8°C if ≤2 days −70°C (dry ice) if >2 days
(Endo)tracheal aspirate, nasopharyngeal or nasal wash/aspirate	Sterile container[a]	2–8°C	2–8°C if ≤2 days −70°C (dry ice) if >2 days
Sputum	Sterile container	2–8°C	2–8°C if ≤2 days −70°C (dry ice) if >2 days
Tissue from biopsy or autopsy including from lung	Sterile container with saline or VTM	2–8°C	2–8°C if ≤24 h −70°C (dry ice) if >24 h
Serum	Serum separator tubes (adults: collect 3–5 mL whole blood)	2–8°C	2–8°C if ≤5 days −70°C (dry ice) if >5 days
Whole blood	Collection tube	2–8°C	2–8°C if ≤5 days −70°C (dry ice) if >5 days
Stool	Stool container	2–8°C	2–8°C if ≤5 days −70°C (dry ice) if >5 days
Urine	Urine collection container	2–8°C	2–8°C if ≤5 days −70°C (dry ice) if >5 days

[a] World Health Organisation, 2020.

for extended periods. In the first part of the day, TB tests are prepared and in the early evening COVID-19 tests. The lab utilizes the Polymerase Chain Reaction (PCR) machine to run COVID-19 samples which requires up 5 h before result is delivered. It is against this background that systems should be incorporated at Sally Mugabe Central Hospital to help improve the rate of COVID-19 testing such as an intelligent robotic testing machine.

1.3.2 COVID-19 effects around the world

Coronaviruses are coursing in creatures and a portion of these viruses have the ability of transmitting among creatures and people which is known as a spillover occasion. Though the novel coronavirus is not deadly as the SARS but the rate of infection seems to be faster than the latter. Fig. 1.2 shows the death rate due to the novel coronavirus as of 13 September 2021 (WHO, 2021; Worldmeter, 2021).

1.3.2.1 Update

There are currently no specific treatments for coronaviruses, but symptoms can be treated (McCreary & Pogue, 2020). Therefore the main focus of this

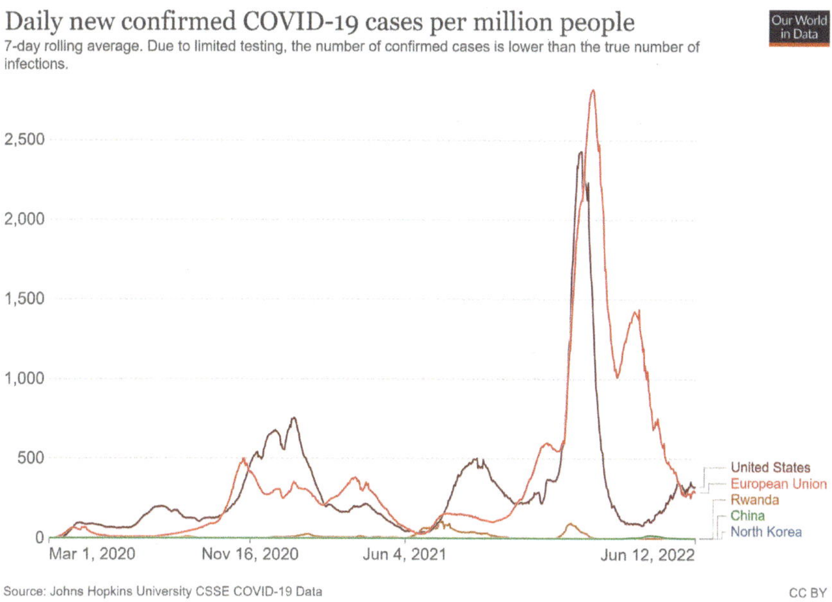

Fig. 1.2 Coronavirus Cases: Trend showing total number of deaths globally. *(Data from Worldmeter. (2021). Covid-19 coronavirus pandemic. Available at: https://www.worldmeter.info/coronavirus/?utm_campaign=homeAdUOA?. (Accessed 22 July 2020).)*

project is to apply artificial intelligence and robotics to automate the virus testing process and help mitigate the spread of the coronavirus. Fig. 1.3 shows the effects of the novel coronavirus to lives of people around the world (UN, 2020).

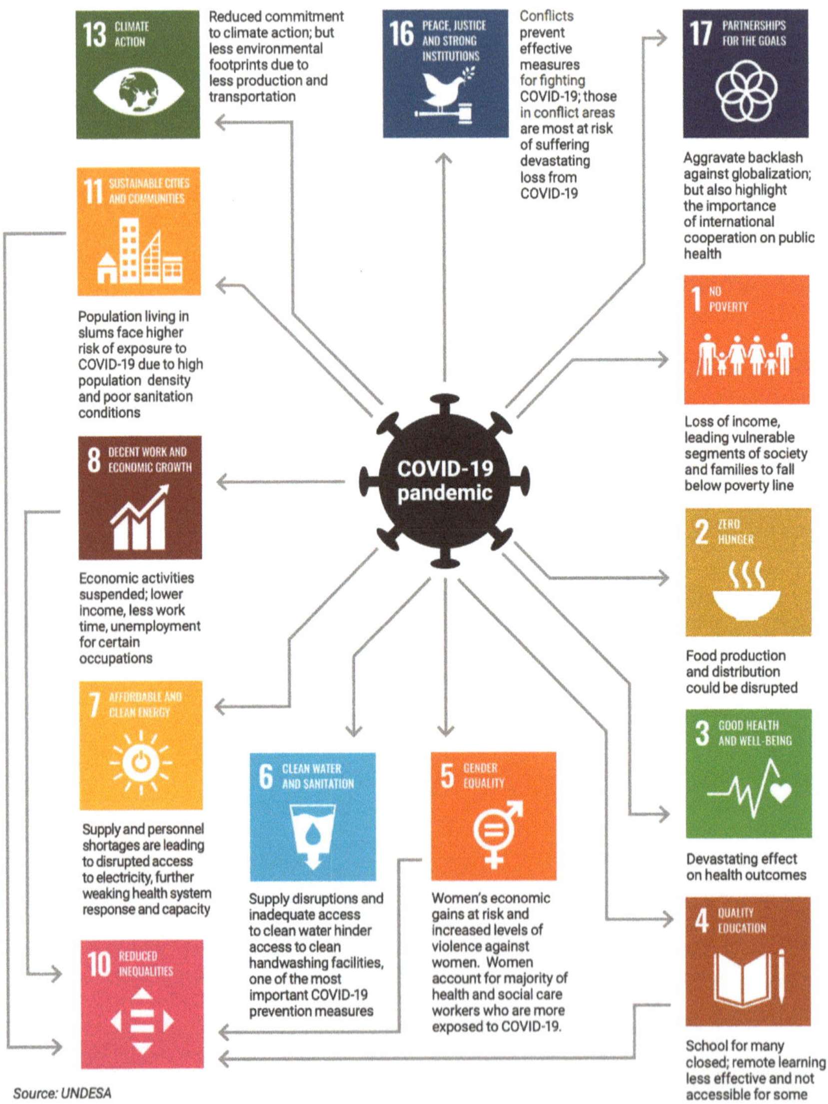

Fig. 1.3 Effects of COVID-19 on the world. *(Data from United Nations, Policy brief series, Summer 2020.)*

1.4 Problem statement

As the search for the cure of the novel coronavirus COVID-19 is on the way, there is still continuous propagation of the infection and the reported total number of deaths is drastically increasing worldwide leading to health services failing to cope with the situation.

1.5 Aim

To develop an appropriate scientific and mathematical model of an intelligent robotic machine for testing COVID-19 patients.

1.6 Objectives

✓ To come up with a robotic manipulator that will be located at the hospital entrance and identifies a human being as he or she comes.
✓ To develop a complete connection of the arm to the entrance architecture.

Using these features, without any human supervision, the robotic manipulator should be able to:

✓ assess the victim's health (checking for COVID-19 symptoms such as high fever)
✓ test patients for COVID-19
✓ provide professional medical practitioners with test results and extra samples for further testing.

1.6.1 Laws of robotics

According to Asimov, there are three laws of robotics and later included the zeroth law (Apoorve, 2015); these include:

Law 0: A robot may not harm humankind or through inaction, permit mankind to come to harm.

Law 1: A robot may not harm a person or through inaction, permit an individual to come to harm, except if this would disregard a higher order law.

Law 2: A robot must obey orders given to it by human beings, except where such orders would conflict with a higher order law.

Law 3: A robot should secure its own reality as long as such assurance does not interfere with a higher order law.

1.7 Justification

1.7.1 Enhance human safety

During excessive strain on the healthcare system such as outbreak of the novel coronavirus pandemic, robotic frameworks can viably decrease the danger of infectious disease transmission to forefront medical services laborers. This can be achieved easily as robotic systems make it achievable to triage, assess, test, and diagnose patients from a distance which is safe for a health worker.

1.7.2 Improve testing rate

Through the use of robots, manual tasks that are labor intensive, time consuming, and repetitive can be handled with ease, thereby reducing the burden on frontline healthcare workers. For example, this will be highly beneficial in lab testing which is currently a very slow process (which might even take days to deliver results to patients).

1.7.3 Practical solution

Due to the continuous rise in COVID-19 cases, most hospitals are having difficulties such as shortage of staff and personal protective equipment. The use of robotics in conjunction with artificial intelligence can help address these concerns with great ease.

1.7.4 Improved accuracy

As already mentioned, robots perform task more rapidly compared to health workers and hence all the different coronavirus (COVID-19) testing techniques can be done for each and every single personnel being tested at a shorter period of time. This will in turn improve the accuracy of the test results since the output will consist of all the technique's strengths.

1.8 Related studies

1.8.1 Drone use during COVID-19 (International Transport Forum, 2020)

The need to reduce spread of the novel COVID-19 virus through minimal human contact (CDC, 2020) has provided a boost to the use of drones. After the outbreak of coronavirus crisis, drones have been used to deliver medical supplies, collect or dispatch of lab samples, monitor social distancing, make public announcements, or disinfect public spaces. For example, China has

Fig. 1.4 Workers in China, refill a drone with disinfectant. *(Data from Murphy, R. R. (2020). Robots are playing many roles in the coronavirus crisis and offering lessons for future disasters. Available at: https://theconversation.com/robots-are-playing-many-roles-in-the-coronavirus-crisis-and-offering-lessons-for-future-disasters-135527. (Accessed 21 July 2020).)*

adapted agricultural drones designed for spraying pesticides in open fields for spraying disinfectants in crowded urban cityscapes (Murphy, 2020) as shown in Fig. 1.4.

1.8.2 Abbott ID NOW

Abbott has designed a new COVID-19 testing machine called ID NOW which can be used to detect the novel coronavirus within a few minutes (Abbott, 2020). The Food and Drug Administration (FDA) has granted Abbott emergency use authorization for this machine. ID NOW provides the fastest molecular test for the novel coronavirus and delivers positive results within 5 min and negative results within 13 min. The ID NOW is a lightweight box (6.6 pound and size of a small toaster) that can be placed in a variety of locations. Due to the portable size of the ID NOW, it can be used in more nontraditional places where people can receive their results in a matter of minutes. The method of operation involves first taking a swab from the nose or back of the throat and mixing it with chemical solutions to release the virus RNA (Cortez, 2020). The mixture is then inserted into the ID NOW system, which has a technology to identify and amplify select sequences of the coronavirus genome and neglects contamination from other viruses (see Fig. 1.5).

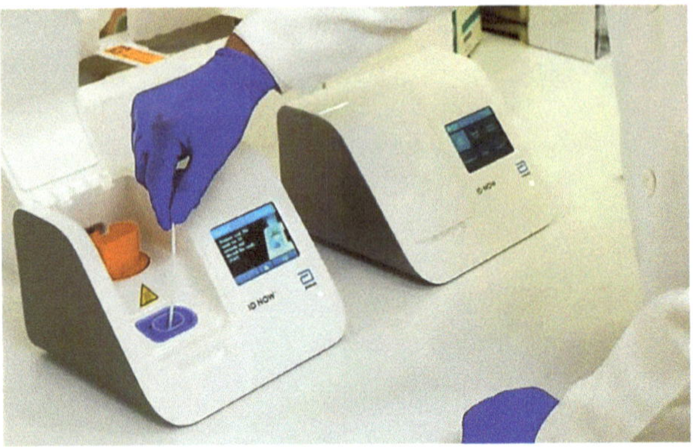

Fig. 1.5 Abbott ID NOW: COVID-19 testing machine. *(Data from Cortez, M. (2020). Virus news: Abbott launches 5-minute Covid-19 test. Bloomberg. Available at: https://www.bloomberg.com/news/articles/2020-03-27/abbott-launches-5-minute-covid-19-test-for-use-almost-anywhere. (Accessed 18 August 2020).)*

1.9 MIT COVID-19 testing trailer

MIT has designed a 60-ft trailer that is now being used as the main testing site for asymptomatic MIT members who need to return to campus (Chu, 2020).

The renovated trailer consists of a check in station and six testing bays. Plastic partitions starting from ceiling to the floor are fixed across the trailer, separating medical staff and those to be tested. Each testing bay has large rubber gloves in which the tester will fit their arms and these will extend to the other side of the partition (see Fig. 1.6). This allows the tester to collect the

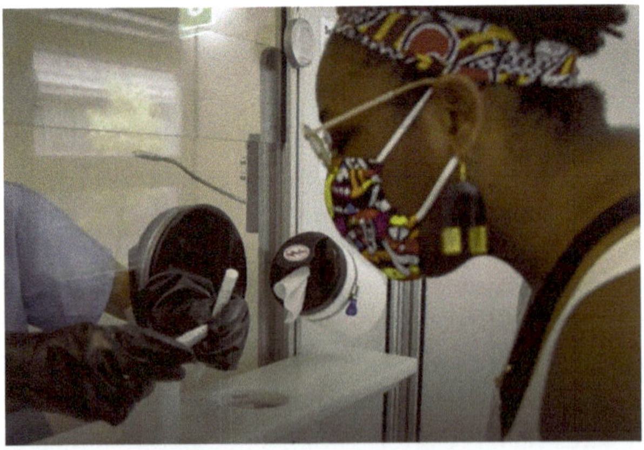

Fig. 1.6 MIT COVID-19 testing in a new designed trailer. *(Data from Chu, J. (2020). How MIT built its own Covid-19 testing trailer. MIT News. Available at: http://news.mit.edu/2020/covid-19-testing-trailer-0813. (Accessed 18 August 2020).)*

nasal swab without either party coming in physical contact. The trailer is also calibrated such that air on either side of the partitions will not mix. The two partitions within the trailer make it possible for the medical staff to safely test people while wearing a surgical mask only, rather than a full Personal Protective Equipment (PPE) (Chu, 2020).

1.10 Conclusion

The World Health Organization (WHO) has termed the novel coronavirus as a worldwide pandemic which will affect every aspect of life as we know it. Thus there is a desperate need to adopt systems which can help the Health Services fight against pandemics such as COVID-19 more effectively and rapidly, to help save more lives. It is against this background that the researchers have proposed the incorporation of an intelligent robotic machine in health services to help test for the COVID-19 virus.

References

Abbott. (2020). *Detect COVID-19 in as little as 5 minutes*. Abbott Newsroom. Available at: https://www.abbott.com/corpnewsroom/product-and-innovation/detect-covid-19-in-as-little-as-5-minutes.html (Accessed 18 August 2020).

Apoorve. (2015). *Industrial manipulators or robotics manipulators*. Elsevier.

Azar, A. T. (2012). *Medical robotics* (May 2014). https://doi.org/10.4018/978-1-4666-4607-0.ch054.

Ben-Ari, M., & Mondada, F. (2018). *Robots and their applications* (pp. 1–20). SpringerLink.

Blacker, P., & Gao, Y. (2018). Robotics and AI-enabled on-orbit operations with future generation of small satellites. *Proceedings of the IEEE*. https://doi.org/10.1109/JPROC.2018.2794829.

Brogardh, T. (2007). Present and future robot control development: An industrial perspective. *Annual Reviews in Control*, *31*(1), 69–79.

CDC. (2020). *How to protect yourself and others know how it spreads*. 2019.

Chu, J. (2020). *How MIT built its own Covid-19 testing trailer*. MIT News. Available at: http://news.mit.edu/2020/covid-19-testing-trailer-0813 (Accessed 18 August 2020).

Cortez, M. (2020). *Virus news: Abbott launches 5-minute Covid-19 test*. Bloomberg. Available at: https://www.bloomberg.com/news/articles/2020-03-27/abbott-launches-5-minute-covid-19-test-for-use-almost-anywhere (Accessed 18 August 2020).

Cross, L., et al. (2009). Best practices and considerations: Including young children with disabilities in early childhood settings. *Focus on Exceptional Children*, *41*(8), 1–8.

DiMaio, S., Hanuschik, M., & Kreaden, U. (2011). The da Vinci surgical system. In J. Rosen (Ed.), *Surgical robotics* (1st ed., pp. 199–217). Springer.

Druce, J., et al. (2011). Evaluation of swabs, transport media, and specimen transport conditions for optimal detection of viruses by PCR. *Journal of Clinical Microbiology*, *50*, 1064–1065.

Dung, N. M., et al. (2007). Two-wheeled welding mobile robot for tracking a smooth curved welding path using adaptive sliding-mode control technique. *International Journal of Control, Automation and Systems*, *5*(3), 283–294. 01.

Hanumara, N. C. (2012). *Efficient design of precision medical robotics*. IEEE.
Huang, C., et al. (2020). Clinical features of patients infected with 2019 novel coronavirus in Wuhan, China. *The Lancet*, 117–131.
International Transport Forum. (2020). *Drones in the era of coronavirus*. (June).
Jayaweera, N., & Webb, P. (2010). Metrology-assisted robotic processing of aerospace applications. *International Journal of Computer Integrated Manufacturing, 23*(01), 283–296.
Kapila, V. (2010). *Introduction to robotics*. Springer.
Kim, H. S., & Song, J. B. (2014). Multi-DOF counterbalance mechanism for a service robot arm. *IEEE/ASME Transactions on Mechatronics, 19*(01), 1756–1763.
Kubatana.net. (2020). *Experts explain Covid-19 testing procedure at Mpilo Hospital*. Available at: http://kubatana.net/2020/04/21/experts-explain-covid-19-testing-procedure-at-mpilo-hospital/.
Kumar, V. (1997). *Introduction to robotics*. What is a robot? History Walking robots The Honda Humanoid The Honda Humanoid What is a robot? The robot is a computer-controlled device that combines the l do anatomy of a robot basic components l ser. (1985).
Letko, M., & Munster, V. (2020). Functional assessment of cell entry and receptor usage for lineage B β-coronaviruses, including 2019-nCoV. *bioRxiv*, 915660.
Mahomed, F. (2021). *Health worker suspensions amplify effects of Zim's Covid-19 second wave*. Available at: https://www.dailymaverick.co.za/article/2021-01-18-health-worker-suspension-amplify-effects-of-zims-covid-19-second-wave/amp/.
McCreary, E. K., & Pogue, J. M. (2020). Coronavirus disease 2019 treatment: A review of early and emerging options. *Open Forum Infectious Diseases*, 7, ofaa105.
Murphy, R. R. (2020). *Robots are playing many roles in the coronavirus crisis and offering lessons for future disasters*. Available at: https://theconversation.com/robots-are-playing-many-roles-in-the-coronavirus-crisis-and-offering-lessons-for-future-disasters-135527 (Accessed 21 July 2020).
Shen, Y., & Li, C. (2020). *Airborne transmission of COVID-19: Epidemiologic evidence from two outbreak investigations* (April). https://doi.org/10.13140/RG.2.2.36685.38881.
Singh, G., Mishra, A., & Sagar, D. (2013). An overview of artificial intelligence. *SBIT Journal of Sciences and Technology*, (1), 3–6.
Soleimani, F., et al. (2011). *Izobraževanje/education robots and medicine—Shaping and defining the future of surgery, endovascular surgery*. IEEE.
Terris, D. J., et al. (2014). A history of robots: From science fiction to surgical robotics a history of robots. *Journal of Robotic Surgery*. https://doi.org/10.1007/s11701-007-0021-2 (July 2007).
UN. (2020). *Shared responsibility, global solidarity: Responding to the socio-economic impacts of COVID-19*. (March).
Wang, C., et al. (2020). A novel coronavirus outbreak of global health concern. *The Lancet*. https://doi.org/10.1002/9781118955567.wbieoc083 (January).
Welch, A. (1995). The self-efficacy of primary teachers in art education. *Issues in Educational Research, 5*(1), 71–78.
WHO. (2020a). *Novel coronavirus (2019-nCoV) situation report—7 2020 [28 January 2020]*. Available at: https://www.who.int/docs/default-source/coronaviruse/situation-reports/20200127sitrep-7-2019- -ncov.pdf?sfvrsn=98ef79f5_2.
WHO. (2020b). *Coronavirus disease 2019 (COVID-19)*. 2019(February).
WHO. (2021). *World health statistics 2020: monitoring health for the SDGs, sustainable development goals*. Available at: https://creativecommons.org/licenses/by-nc-sa/3.0/igo.
Wilcox, B., Ambrose, R., & Kumar, V. (2008). Exploratory investigation of attitudes towards assistive robots for future users (Chapter 3), pp. 25–39.
World Health Organisation. (2018). *Protocol to investigate non-seasonal influenza and other emerging acute respiratory diseases*. Geneva: World Health Organization. Available at: https://www.who.int/influenza/resources/publications/outbreak_investigation_protocol/en/.

World Health Organisation. (2020). *Laboratory biosafety guidance related to the novel coronavirus (2019-nCoV).*

World Health Organization. (2020a). *Covid-19 strategy up date.* (April).

World Health Organization. (2020b). *Infection prevention and control during health care when novel coronavirus (nCoV) infection is suspected, interim guidance, January 2020.* Geneva: World Health Organization. Available at: https://www.who.int/publicationsdetail/infection-prevention-and-controlduring-health-care-when-novelcoronavirus-(ncov)-infection-is-suspected2020012.

World Health Organization. (2020c). *Laboratory testing for coronavirus disease (COVID-19) in suspected human cases* (March) (pp. 1–7).

World Health Organization. (2020d). *WHO statement regarding cluster of pneumonia cases in Wuhan, China Geneva 2020 [updated 9 January 2020 and 14 January 2020].* (January). Available at: https://www.who.int/china/news/detail/09-01-2020-who-statement-regarding-cluster-of-pneumoniacases-in-wuhan-china.

Worldmeter. (2021). *Covid-19 coronavirus pandemic.* Available at: https://www.worldmeter.info/coronavirus/?utm_campaign=homeAdUOA (Accessed 22 July 2020).

Xu, K., Cai, H., Shen, Y., Ni, Q., Chen, Y., Hu, S., et al. (2020). Management of COVID-19: the Zhejiang experience. *Journal of Zhejiang University.* https://doi.org/10.3785/j.issn.1008-9292.2020.02.02.

Zhang, Y., Chen, C., Zhu, S., Shu, C., Wang, D., Song, J., et al. (2020). Isolation of 2019-nCoV from a stool specimen of a laboratory-confirmed case of the coronavirus disease 2019 (COVID-19). *China CDC weekly, 2*(8).

Zhu, N., et al. (2020). A Novel coronavirus from patients with Pneumonia in China, 2019. *New England Journal of Medicine*, 1–2.

CHAPTER 2

Uses of artificial intelligence and robotics in healthcare systems

2.1 Introduction

From generation to generation, human safety has always come first above all other factors that affect human live. In regard to this, a lot of effort is always put in research and developments of techniques that eliminate any threats to the existence of the human race (or any species for that matter). However, the novel coronavirus COVID-19 has brought sense of insecurity to lives of people as it continues to be one of the leading causes of death around the globe. Systems and laws have been adapted as mitigation measures to the spread of this virus; though this has shown good results still more people are being affected and the death toll increases every day. The World Health Organization recorded 2977 deaths as of 29 February 2020 and 89,411 deaths as of 9 April 2020 (WHO, 2020). However, all is not lost, as the idea of robotics and AI can help overcome these challenges without having to pull a muscle that much. For example, robot-assisted surgery was adapted in the early 1980s and still continues to grow in functionality and help save more lives every year. This shows that the same principles can be modified to assist and manage the recent global pandemic COVID-19 virus.

2.2 Robotic manipulator parts and components

Generally a robot can be characterized as a machine intended to perform one or more automatic tasks precise and rapidly (Rouse, 2020). Some medical robots can be regulated remotely by a human administrator and the range can be short to very long distances while other robots are autonomous (Fijany & Bejczy, 1993; Khatib, 2019; Russo et al., 2019). An autonomous robot executes tasks independently of human interaction (Brogardh, 2007). Therefore it is a standalone system with its own computer (controller). For example, an advanced smart robot has a built-in artificial intelligent system that is capable of learning from past experience and the environment it is exposed to (Brogardh, 2007; Dung et al., 2007; Jayaweera & Webb,

Healthcare Systems Design of Intelligent Testing Centers
https://doi.org/10.1016/B978-0-323-99443-9.00010-3

2010; Kim & Song, 2014). Robots are usually grouped according to the time frame which they were mostly used. First-generation robots include stationary, nonprogrammable, electrochemical devices without sensors and were widely used in the 1970s. Second-generation robots consist of sensors and programmable controllers and were designed in the 1980s. Third-generation robots contain sophisticated programming, speech recognition, and other advanced features and they were developed in the 1990s and the present. Fourth-generation robots include mainly artificial intelligence, self-assembly, and nanoscale size and they are currently in research and development phase.

In any programmable medical robot, there will always be one of the following (see Fig. 2.1 and Table 2.1) (Samiksha, 2020).

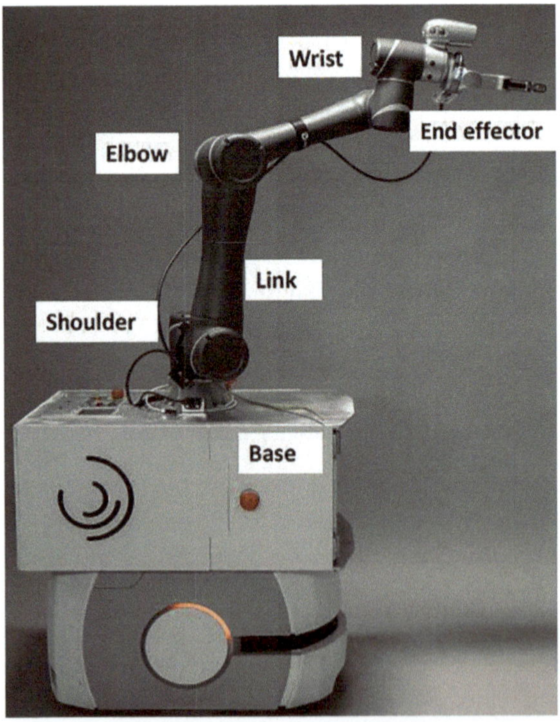

Fig. 2.1 General anatomy of a robotic manipulator. *(Courtesy Danilyuk, P. (2021).* Black and silver steel tool. *Free Stock Photo. Available at: https://www.pexels.com/photo/black-and-silver-steel-tool-8438864/. (Accessed 14 September 2021).)*

Table 2.1 Robotics parts and components.

Robot component	Description
1. Manipulator	A manipulator is a robotic arm like that of a human being, having several joints and connections. The manipulator can be isolated into two sections, each with its own unique functions and these are wrist and arm and body (AZoRobotics, 2013). The manipulator of the robot comprises of joints associated by links. They can be utilized to move the end effector to the desired position within a given workspace. The function of the wrist is to be attached to the end effector which interacts with the surroundings
2. End effector	The manipulator is usually fixed at the base and the other end attached to an end effector (Bouchard, 2014; Kaur, 2013b). The end effector is a member of the robot that interacts with the surrounding environment
3. The locomotion device	The locomotion of the robot is provided by actuators. The types of actuators used depend on the source of energy used, for example electric actuators are electrically powered
4. The controller	A robot consists of a controller which is analogous to the human brain (Collett et al., 2005; de Souza Jacob et al., 2012; Kiesler et al., 2008). Through the help of the controller, robots can carry out the desired tasks
5. The sensors	One of the most important parts of a robot is the sensors. Sensors are used to gauge quantities like position, speed, force, proximity, torque, temperature, and these are fed to the controller which in turn controls the robot

2.3 Medical robot's manipulator configurations

There are a wide range of robots utilized in medical field and they are the same types used in present-day businesses and working environments, only the application of the robot will be different. Some of the main types include revolute or articulate, cylindrical, polar, and SCARA robots.

2.3.1 Articulated or revolute geometry arm

At the point when the elbow joint of the Cartesian arm is replaced with the revolute joint, an articulated manipulator is created that works in an unpredictable thick-walled circular shell (Kah, Shrestha, & Hiltunen, 2015) (see Fig. 2.2).

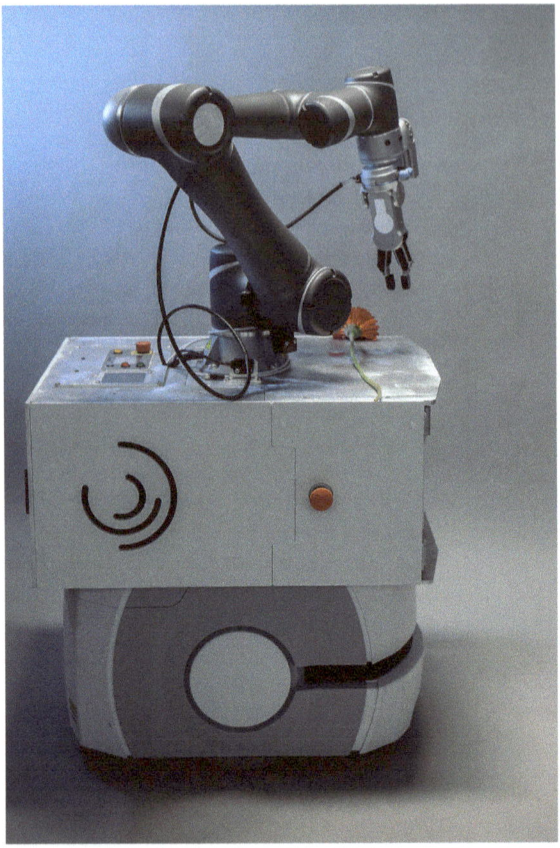

Fig. 2.2 Articulated or revolute geometry arm. *(Courtesy Danilyuk, P. (2021).* Man in white button up shirt holding black and gray camera. *Free Stock Photo. Available at: https://www.pexels.com/photo/man-in-white-button-up-shirt-holding-black-and-gray-camera-8439008/. (Accessed 14 September 2021).)*

2.3.2 Cartesian robots

Cartesian robots operate based on the Cartesian coordinate system (RS, 2020). Therefore their movement is controlled based on the x, y, and z axes (Machine Design, 2020). They are mostly used for performing errands like picking and setting objects as well as machine tooling. The mode of operation is achieved through the use of linear actuators. For instance, the Cartesian robot can situate an apparatus in three-dimensional space and control it through a progression of linear motions to switch between positions. Fig. 2.3 shows an example of a Cartesian robot.

Fig. 2.3 Cartesian robot example. *(Courtesy Machine Design. (2020).* The difference between cartesian, six-axis, and SCARA robots. *Machine Design. Available at: https://www.machinedesign.com/mechanical-motion-systems/article/21831692/the-difference-between-cartesian-sixaxis-and-scara-robots. (Accessed 14 August 2020).)*

2.3.3 Cylindrical robots

Cylindrical robots (see Fig. 2.4) are programmed so that they can play out their errands inside a barrel-shaped work envelope (Sitemap, 2007). The work envelope is round and hollow in shape, that is, the robot can perform movements up, down, and around a focal axis. This can be accomplished using both revolving and direct actuators, enabling the robot to have both straight and rotational movement (RS, 2020). The utilizations of cylindrical robots incorporate spot welding and machine device handling.

2.3.4 Polar robots

A polar robot is a complex sort of robot that works inside a circular work envelope (RS, 2020). This can be achieved through the combination of two turning joints and a direct joint. Also known as spherical robot, this robot can perform similar tasks as the cylindrical robot but with an expanded capacity to move to and from inside the work space. Therefore the applications incorporate handling machine devices, spot and arc welding, and die casting.

2.3.5 SCARA robots

SCARA is an abbreviation that represents Selective Compliance Assembly Robot Arm (or sometimes Selective Compliance Articulated Robot Arm) (Epson, 2020). Consistence alludes to a level of suitable adaptability in development with regard to robotics. SCARA robots are designed so that they

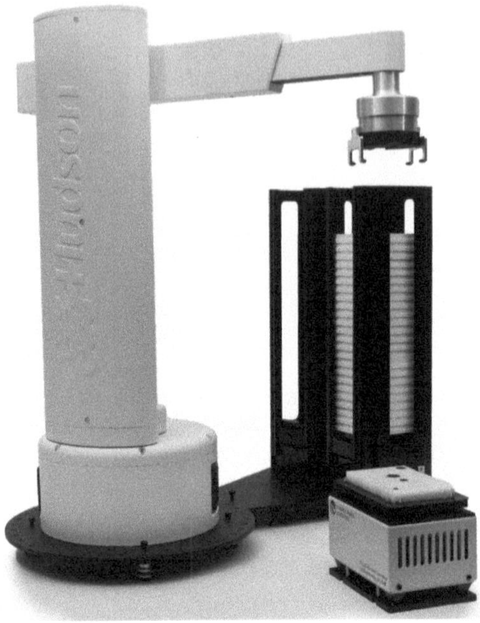

Fig. 2.4 Cylindrical robot example. *(Courtesy Sitemap. (2007).* Cylindrical robots. *Available at: https://www.industrial-electronics.com/industrial-electricity-com/4_Cylindrical_Robots.html. (Accessed 14 August 2020).)*

permit just a little level of motion in a couple of directions while fixed in the excess axes. The SCARA robot (see Fig. 2.5) is most appropriate for certain assembly or part placement assignments.

2.4 End effector design

An end effector is the gadget at the robotic manipulator end that is intended to collaborate with the environment (Bouchard, 2014). The kind of the end effector relies upon the utilization of the robot. Therefore the end effector is the last connection of the robotic arm. Some of end effectors include:

✓ Mechanical grippers
✓ Negative pressing factor (vacuum) tools
✓ Magnetic tools
✓ Hooks
✓ Ladles (scoops liquid or powder)
✓ Electrostatic sprays

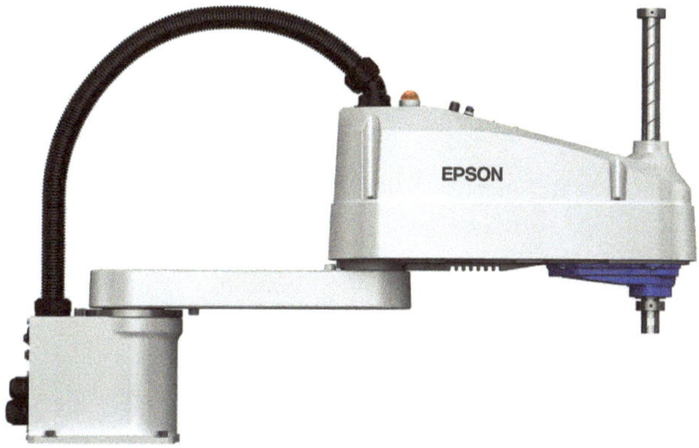

Fig. 2.5 Example of a SCARA robot. *(Courtesy Epson. (2020).* Epson SCARA LS10-B702S (standard). *Epson. Available at: https://www.epson.eu/products/robot/epson-scara-ls10-b702s-standard. (Accessed 14 August 2020).)*

2.4.1 Mechanical grippers

The earlier gripper (see Fig. 2.6) is an illustration of an electric gripper end effector. Grippers are the mostly used types of robotic manipulators (Bolee Machine Tool, 2019). Depending on the application, they can utilize distinctive holding techniques and actuation styles. The mechanical gripper may have two, three, four, or even five fingers. Four general classes of mechanical grippers include:

✓ Impactive—hooks or jaws that grip an object straightforwardly through an actual impact.
✓ Ingressive—pins, needles, pins, or hackles that enter physically through the object surface.
✓ Astrictive—attraction forces applied to the outside of the item by magneto, electro, or vacuum bond.
✓ Contigutive—requires direct contact to empower attachment like paste, surface pressure.

2.5 Artificial intelligence and robotics in the medical field

AI is getting increasingly sophisticated in performing tasks that previously only human being can do but more efficiently at a lower cost (Murphy, 2020). One of the artificial intelligence's potential advantages is in assisting individuals with remaining well-being so they do not bother with physicians frequently. For instance, the utilization of artificial intelligence and the Internet of Medical Things (IoMT) in customer medical applications is as

Fig. 2.6 Two-finger gripper. *(Courtesy Danilyuk, P. (2021).* Black and silver steel tool. *Free Stock Photo. Available at: https://www.pexels.com/photo/black-and-silver-steel-tool-8438864/. (Accessed 14 September 2021).)*

of now helping many individuals (Dautenhahn et al., 2006; Tiwari et al., 2010). What's more, artificial intelligence is being utilized to distinguish infections, like cancer, all the more precisely and in their beginning phases (Tavakoli, Carriere, & Torabi, 2020). Therefore several artificial intelligence strategies, such as machine learning, internet of things, fuzzy logic, and genetic algorithms (Hong, Nakhaeinia, & Karasfi, 2012), can help improve the performance of medical robots (Fig. 2.7).

2.5.1 Internet of Things (IoT)

Internet of Things (IoT) is an arrangement of interrelated computer gadgets, mechanical and advanced machines, objects, people, or creatures that have exceptional identifiers and the capacity to move information through an

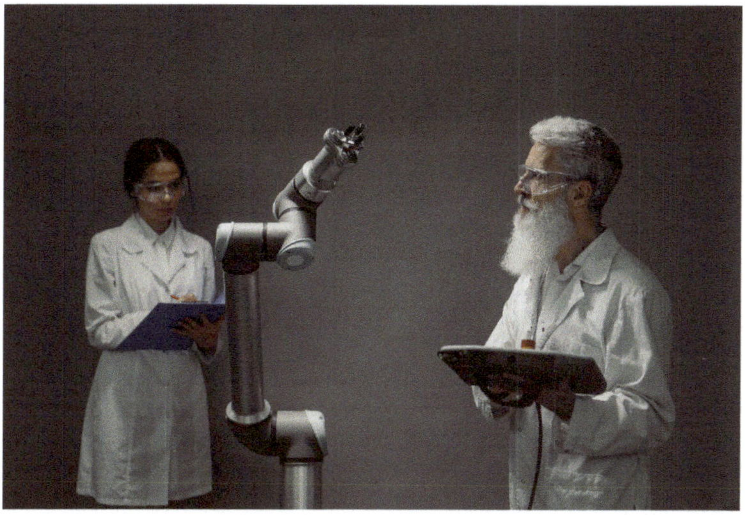

Fig. 2.7 A depiction of the applications of robotics and AI in the medical field. *(Courtesy Danilyuk, P. (2021).* Man in white button up shirt holding black and gray camera. *Free Stock Photo. Available at: https://www.pexels.com/photo/man-in-white-button-up-shirt-holding-black-and-gray-camera-8439008/. (Accessed 14 September 2021).)*

organization without expecting human-to-human or human-to-computer interaction (McClelland, 2020). This is a technical definition of IoT, but in short, Internet of Things involves extending the power of internet beyond computers and smartphones to a variety of other things, environment, and processes. Therefore the concept involves connecting everything in the world to the internet. Things that are connected to the internet are able to send or receive information or both. To be able to send and or receive information makes things smart and thus the main benefit of IoT (Burgess, 2018). Smart thing does not need a supercomputer inside them but all the thing has to do is connect to a supercomputer. There are three categories involved in the internet of things, namely, things that collect and send information (sensors), things that receive and act on information, and things that do both, which is where the power of IoT lies (Burgess, 2018; Khan, 2016; McClelland, 2020) (Fig. 2.8).

2.5.2 Fuzzy logic

Reliability and robustness of any medical robot in a dynamic or obscure environment depend on capacity of the robot to deter obscurity without impact and respond fast to vulnerabilities. To foster these errands a

Fig. 2.8 Internet of Things. *(Courtesy Simplilearn (2020).* IoT explained: What it is, how it works, why it matters. *Available at: https://www.simplilearn.com/what-is-iot-how-and-why-it-matters-article. (Accessed 14 August 2020).)*

procedure that utilizes human thinking and decision making must be adapted. Thus, fuzzy logic comes in (GeeksforGeeks, 2019), it gives an alternative obtain the expertise of the human brain. The concept of the Fuzzy Logic Controller (FLC) is essential in solving in the reality uncertainties and the precise model of the environment is usually not absolutely needed for testing (Hong et al., 2012). Therefore a simple design can be used to come up with a very robust and intelligent medical robot.

2.5.3 Application of robots in the medical field

Robots have been adapted in every aspect of life from science fables to public hospitals, where they are evolving healthcare. Some applications of robotics and artificial intelligence in healthcare are summarized as follows.

2.5.3.1 Robots for surgical precision

Surgery is an experience that is quite unpleasant and usually takes a lot of time (Hanumara, 2012). Depending on the man power available, the list to undergo any type of surgery is usually very long. Thus surgical robots are very essential in medical healthcare as they improve the efficiency of surgery (Soleimani et al., 2011).

One of the most common surgical robots is the da Vinci Surgical System (see earlier Fig. 2.9) (DiMaio, Hanuschik, & Kreaden, 2011). The main features of this robotic framework are a magnified three-dimensional high-definition vision framework and tiny wristed apparatus that is more flexible than the human hand (see Fig. 2.9). The surgeon has 100% regulation of the

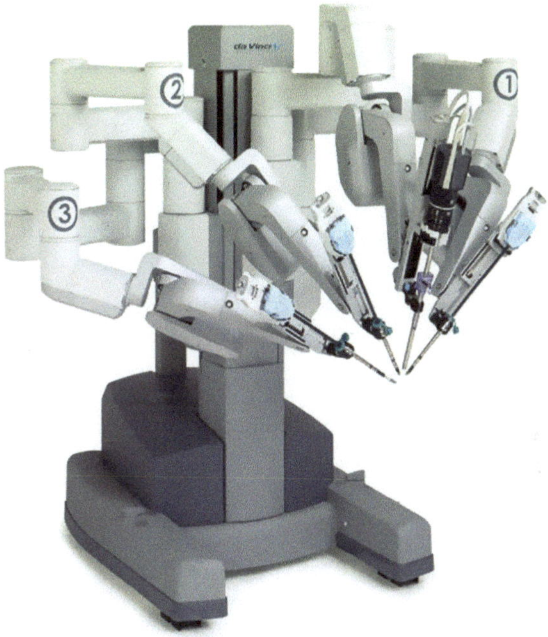

Fig. 2.9 da Vinci surgical robotic system. *(Courtesy Desai, J. P., et al. (2018). The encyclopedia of medical robotics.* World Scientific. *https://doi.org/10.1142/10770-vol1.)*

robotic framework at all the time and hence, high precision is achieved (Adascalitei, Asachi, & Adăscăliţei, 2011; Lavallee et al., 1992; Taylor & Stoianovici, 2003) by combining the strength of the robotic machine and surgeon (Berndsen & Luckner, 2012; Tsui & Yanco, 2007).

2.5.3.2 Handling repetitive tasks

Any patient in a hospital interacts more with nurses. Nurses are usually there to check vital signs, monitor the patient condition, and take care of their hygiene if needed. These tasks are usually daunting, leading to the experience being unpleasant for everyone involved. Thus robotic nurses have to be incorporated (The Medical Futurist, 2020). These machines are designed to perform repetitive and tedious errands and humans can focus more on issues that require human decision-making skills, care, and empathy (Owen-Hill, 2019). The TUG autonomous mobile delivery robot (Businesswire, 2014) is an example of such a robotic framework. This robot is able to carry carts or bins with a net mass of up 453 kg, consisting mainly of medications and laboratory samples (see Fig. 2.10).

Fig. 2.10 An example of an autonomous mobile delivery robot. *(Courtesy Kindel Media. (2021).* Delivery robots parked beside building. *Free Stock Photo. Available at: https://www.pexels.com/photo/delivery-robots-parked-beside-building-8566535/. (Accessed 14 September 2021).)*

2.5.3.3 Telemedicine

In some cases, accidents happen in places such that medical professionals would not be able to reach victims in time (Healthcare Administration Degree Programs, 2020; The Medical Futurist, 2020). Systems such as Health Tap have been developed so that patients in far-off areas have access to excellent emergency consultations for medical emergencies such as stroke. This is done through the use of smartphones and computers (Pande et al., 2003) and the process is called telemedicine (Crawford, 2016).

2.5.3.4 Disinfecting hospitals

According to the CDC statistics used by Xenex, hospital-acquired infections (HAI) are one of the main sources of death in the United States (The Medical Futurist, 2020). In addition, to loss in human life, these infections are costly to hospitals. Thus the Xenex robot was designed that uses high-intensity ultraviolet light to disinfect healthcare system more efficiently (see Fig. 2.11).

2.5.3.5 Nanorobots swimming in blood

Scientists from the Max Planck Institute have been testing extraordinary micro-sized robots (Durairaj & Shanker, 2012) that deliver drugs and other medical relief to specific location by swimming through the body fluids. For example, the origami robot (MIT, 2015) (see Fig. 2.12) contains a capsule that will dissolve in the stomach once it is swallowed and then it unfolds itself (Burgner-Kahrs, Rucker, & Choset, 2015; Dario et al., 1996; Sivasankar & Durairaj, 2012). The robot is used to patch up wounds in the stomach guided by a technician with help of the magnetic field.

2.6 Steps in design and development of robots

A robot is a complex system and requires an understanding of electricity, mechanical design, and programming to do work. The following are the elements of robotics (gagileo.org, 2020; RS, 2020).

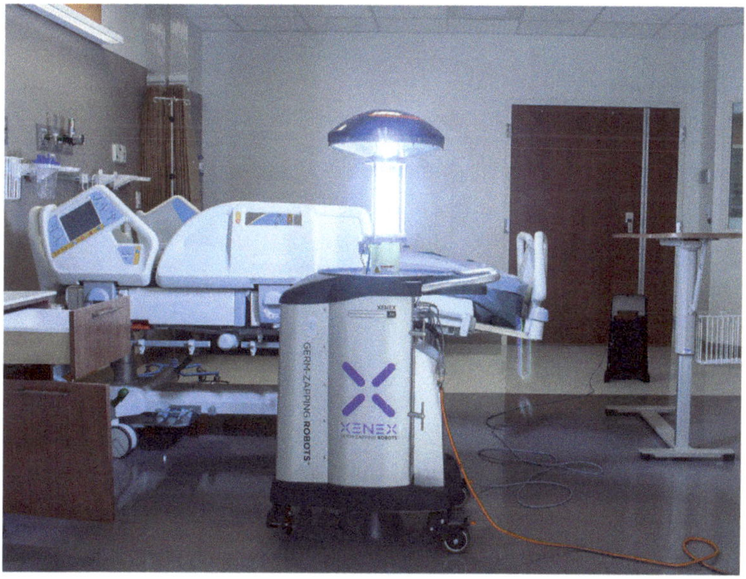

Fig. 2.11 Xenex U-V disinfecting robot. *(Courtesy Begić, A. (2018). Application of service robots for disinfection in medical institutions. In* Lecture notes in networks and systems *(pp. 1056–1065). Springer. https://doi.org/10.1007/978-3-319-71321-2_89.)*

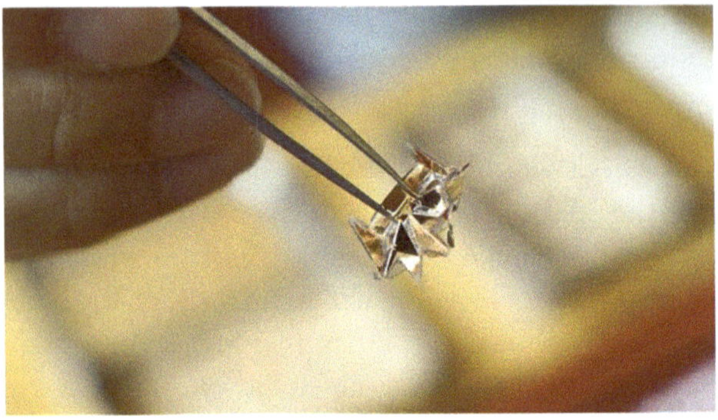

Fig. 2.12 Origami robot. *(Courtesy MIT. (2015).* Robot origami: Robot self-folds, walks, and completes tasks. *YouTube. Available at: https://www.youtube.com/watch?v=ZVYz7g-qLjs. (Accessed 14 August 2020).)*

2.6.1 Mechanical design

The essential part of making a robot is to make it move and this is where mechanical engineering is applied (Alp, 2002; Wolf et al., 2005; Yoshikawa, 1985). Mechanical engineering focuses more on machine design but robotics design will focus more on motors and gears and how to manipulate them to get the desired motion of the robot (Al-Shuka & Song, 2019; Khatib, 2019; Russo et al., 2019).

2.6.2 Electronics

Another fundamental part of robotics is circuit designing, that is, electronics. An understanding of electronics and microcontrollers (such as Raspberry Pi) is essential for selecting right components of the robot being designed (Rodríguez León et al., 2019; Russo et al., 2019).

2.6.3 Programming

After the mechanical and electrical design, then the robot is ready to be programmed. The microcontroller which is the brain of the robot is the one which will be programmed and this is crucial because the robot needs instruction on how to interact with the surrounding environment.

2.6.4 Fabrication

The final stage in building a robot involves fabrication, where all the components are assembled together in the robotic structure (Kim & Wensing, 2017). This will be done with the help of fabrication tools, such as laser cutters, Computer Numerically Controlled (CNC) mills, and 3D printers. The materials chosen for the robotic structure will depend on the application of the robot being designed.

2.7 Power supply of robots

Power sources are useful while designing robotic frameworks. Therefore the determination of power sources should be one of the essential spotlights due to its impact on the mechanism, bundling, weight, and size of the robotic framework being designed. Some of the power sources used in robotics are listed as follows (Kaur, 2013a).

2.7.1 Generator systems

A generator (see following figure) operates based on conversion of rotational mechanical energy produced by a combustion engine to electrical energy which will be used to power the robot. This is two-stage system which can charge a robot consistently during activity, in any event, when the robots are stationary.

2.7.2 Batteries

Batteries are the mainly used power system for robotic components. Batteries can be grouped into two categories: rechargeable or nonrechargeable batteries (Wright, 2020). Nonrechargeable batteries usually convey more power depending on their size. Examples of nonrechargeable batteries are alkaline and lithium batteries (see Fig. 2.13).

Rechargeable batteries, on the other hand, deliver small voltage and they are found in battery packs along with specialized battery chargers. Some examples include lead-acid, lithium-ion, and nickel-cadmium batteries.

2.7.3 Photovoltaic cells

Photovoltaic or solar cells are normally utilized to charge batteries of robotic framework. They are utilized together with a capacitor and have the ability to be charged up to a desired voltage value and then be discharged through the motions of a motor.

Fig. 2.13 Alkaline batteries. *(Courtesy Abdelg, M. (2020).* Free stock photo of addition, alkaline, battery. *Available at: https://www.pexels.com/photo/industry-technology-power-electricity-6015268/. (Accessed 14 September 2021).)*

2.8 Degree of autonomy of robots

Not all robots are the same; there are different levels of autonomy for different types of robots (451 Research, 2019; Bajracharya et al., 2008; Tahboub, 1999; Tahboub et al., 1999). The levels of autonomy of robots are defined as follows:

2.8.1 Level 0

The real industrial robots are at level 0 and this also includes some computer numerically controlled (CNC) instruments. These are placed at 0 to indicate the initial point toward autonomous robotics.

2.8.2 Level 1

This level involves remote regulated gadgets, including bomb disposal robots that give a level of telepresence, protecting administrators from harm, and piloted drones. The robots are regulated using motion translation gadgets such as joy sticks. Therefore this is the anchor point of IoT gadget interactions.

2.8.3 Level 2

Level 2 incorporates smart instruments and sophisticated gadgets that help people directly and where the degree of human interaction includes gesture, voice, and body movement. Generally, this degree involves more of the exoskeleton systems (Nunuparov et al., 2019; Zielińska & Wang, 2019) but also advanced telepresence control systems which use many human gestures to interact.

2.8.4 Level 3

Level 3 involves increased level of autonomy of industrial robots in level 0, that is, the robots now have more processing power and a degree of sense to their surrounding environment.

2.8.5 Level 4

Level 4 involves mesh robotics, which are interconnected gadgets that utilize advanced function code and machine learning to connect with one another to perform undertakings with high changeability in regions like picking and assembly.

2.8.6 Level 5

Level five presents Cobots which are an arising portion of self-governing robots that are intended to work with people but in such a way that helps and comprehends the job that they are performing.

2.8.7 Level 6

Level 6 robots are the same as the Society of Automotive Engineers (SAE) level 5 self-sufficient vehicles (Dremali, 2018). The level of self-sufficiency in these self-driving vehicles requires no human information. The vehicle manages a known undertaking (transport) in a complex yet known space (the road framework). This also includes drones used in delivery and inspection.

2.8.8 Level 7

These robots include regions which were recently held for science fiction, yet they are beginning to show up in exhibitions and recordings. This level includes robots that are completely independent, self-perambulate, and utilize an assortment of approaches to cross the world and engage with assignments. They are still in the examination stage yet are beginning to advance out into the world.

2.8.9 Level 8

Level 8 robots expand the development and route capacities of the level 7 robots, yet they are beginning to perform a wide range of undertakings and change in accordance with the circumstance or job they are given.

2.8.10 Level 9

Level 9 addresses robots that are self-governing, profoundly portable and flexible that work in collaboration with others, consolidating components of Level 4 robot-to-robot correspondence and Level 5 human participation as a cobot.

2.9 Actuators used in robotics

The basic types of robotic motors used are presented in Table 2.2 (CLR, 2020).

Table 2.2 Types of actuators used in robotics.

Type of actuator	Description
AC or DC motors	AC motors are valuable for high torque while DC motors for high velocity applications yet DC motors, specifically, are not typically durable except if altered with extra gears
Servo motors	Servo motor includes an AC, DC, or brushless DC motor joined with a position-detecting gadget and a control circuit. This engine permits exact control of angular position and it is a rotating actuator. The activity is dependent on the utilization of pinion wheels and sensors
Stepper motors	A stepper motor is a brushless DC electric motor that partitions full turn into various equivalent advances. The situation of the motor would then be able to be made to move and hold at one of these steps with any position sensor for input.
Linear actuators	Linear actuators develop straight movement and are intended to withstand a wide scope of working conditions. These actuators work by changing over energy created by an electric motor into torque, which then, at that point, moves a system. The actuator can be controlled mechanically, electrically, or through programming

2.10 Sensors used in medical robots

There are wide robotic sensors used for medical applications and these are summarized in the Table 2.3.

2.10.1 Force-torque sensors

Force-Torque (FT) sensors are pucks positioned between the robot flange and the end effector that interact with the environment (for example a part) (Medical Design Briefs, 2013). They are used to gauge the force and torque that the robot applies to the environment through the end effector. The mode of operation involves measuring the force and torque in the *x*, *y*, and *z* axes. These FT sensors are mainly utilized when the force applied by the robot needs to be controlled (Song & Lee, 1999).

2.11 Conclusion

The research conducted so far confirms that adopting robotics and AI in the healthcare is very crucial for the battle against the novel COVID-19 pandemic. Current testing systems for COVID-19 virus testing have limitations, which may also affect this research. The robotic systems used in medicine and the testing process of COVID-19 virus may be challenging and complex but they can be implemented, as there is a desperate need to stop this global pandemic COVID-19.

Table 2.3 Sensors used in robotics.

Type of sensors	Example
Contact sensors	✓ Push fastens and contact switches ✓ Pressure pads and sensors
Distance sensors and gauges	✓ Ultrasonic range obtainers ✓ Infrared sensors ✓ Laser quantifying gadgets and sensors ✓ Stretch and bend sensors
Positioning sensors	✓ Indoor navigation (room localization) sensors ✓ GPS and other real-time tracking devices
Rotation sensors	✓ Potentiometers ✓ Gyroscopic gadgets
Environmental sensors	✓ Photoelectric or light sensors ✓ Sound sensors ✓ Thermal sensors and thermal imaging cameras ✓ Humidity and moisture sensors ✓ Pressure sensors ✓ Gas sensors

References

451 Research. (2019). *451 Research—Autonomy-level classification for robots in an IIoT World*. Available at: https://go.451research.com/MI-Robots-in-IIoT-World.html (Accessed 24 July 2020).

Adascalitei, F., Asachi, G., & Adăscăliței, F. (2011). *Practical applications for mobile robots based on Mecanum wheels—A systematic survey the Romanian review precision mechanics, optics & mechatronics*. Available at: https://www.researchgate.net/publication/233867057 (Accessed 24 July 2020).

Alp, A. (2002). *Cable suspended robots: Design, planning and control*. ieeexplore.ieee.org. Available at: https://ieeexplore.ieee.org/abstract/document/1014428/ (Accessed 24 July 2020).

Al-Shuka, H. F. N., & Song, R. (2019). Decentralized adaptive partitioned approximation control of robotic manipulators. In *CISM international centre for mechanical sciences, courses and lectures* (pp. 10–17). Springer International Publishing. https://doi.org/10.1007/978-3-319-78963-7_3.

AZoRobotics. (2013). *What is a robotic manipulator?*. Available at: https://www.azorobotics.com/Article.aspx?ArticleID=138 (Accessed 24 July 2020).

Bajracharya, M., et al. (2008). *Autonomy for mars rovers: Past, present, and future*. ieeexplore.ieee.org. Available at: https://ieeexplore.ieee.org/abstract/document/4712499/ (Accessed 24 July 2020).

Berndsen, M., & Luckner, J. (2012). Supporting students who are deaf or hard of hearing in general education classrooms: A Washington state case study. *Communication Disorders Quarterly*, *33*(2), 111–118. https://doi.org/10.1177/152574011038439', in. 10.1007/978-3-319-32552-1_63.

Bolee Machine Tool. (2019). *Robotiq 2-FINGER 85/140 Robot Gripper*. Bolee Machine Tool Ltd. Available at: http://www.bolee.com.hk/robotiq-2-finger-85-robot-gripper/ (Accessed 14 August 2020).

Bouchard, S. (2014). *Robot end effector: Definition and examples*. Available at: https://blog.robotiq.com/bid/53266/Robot-End-Effector-Definition-and-Examples (Accessed 24 July 2020).

Brogardh, T. (2007). Present and future robot control development: An industrial perspective. *Annual Reviews in Control*, *31*(1), 69–79.

Burgess, M. (2018). *What is the internet of things? WIRED explains*. WIRED UK. Available at: https://www.wired.co.uk/article/internet-of-things-what-is-explained-iot (Accessed 14 August 2020).

Burgner-Kahrs, J., Rucker, D. C., & Choset, H. (2015). Continuum robots for medical applications: A survey. *IEEE Transactions on Robotics*, *31*(6), 1261. https://doi.org/10.1109/TRO.2015.2489500.

Businesswire. (2014). *Aethon installs the TUG® autonomous smart mobile robot at 21 hospitals in 2013*. Business Wire. Available at: https://www.businesswire.com/news/home/20140114005110/en/Aethon-Installs-TUG%C2%AE-Autonomous-Smart-Mobile-Robot (Accessed 14 August 2020).

CLR. (2020). *Electrical actuators in robotics: Some amazing innovations*. Blog CLR. Available at: https://clr.es/blog/en/electrical-actuators-robotics/ (Accessed 24 July 2020).

Collett, T., et al. (2005). *Player 2.0: Toward a practical robot programming framework*. leek.ai.sri.com. Available at: http://leek.ai.sri.com/~gerkey/papers/acra2005.pdf (Accessed 24 July 2020).

Crawford, M. (2016). *Top 6 robotic applications in medicine*. ASME. Available at: https://www.asme.org/topics-resources/content/top-6-robotic-applications-in-medicine (Accessed 24 July 2020).

Dario, P., et al. (1996). *Robotics for medical applications*. ieeexplore.ieee.org. Available at: https://ieeexplore.ieee.org/abstract/document/540149/ (Accessed 24 July 2020).

Dautenhahn, K., et al. (2006). *How may I serve you? A robot companion approaching a seated person in a helping context*. dl.acm.org. Available at: https://dl.acm.org/doi/abs/10.1145/1121241.1121272 (Accessed 24 July 2020).

de Souza Jacob, R. T., et al. (2012). Frequency modulation systems in hearing impaired children: Outcome evaluation. *Revista Da Sociedade Brasileira De Fonoaudiologia*, *17*(4), 417–421.

DiMaio, S., Hanuschik, M., & Kreaden, U. (2011). The da Vinci surgical system. In J. Rosen (Ed.), *Surgical robotics* (1st ed., pp. 199–217). Springer.

Dremali, A. (2018). *The 5 levels of autonomy*. Available at: https://www.andplus.com/blog/the-5-levels-of-autonomy (Accessed 24 July 2020).

Dung, N. M., et al. (2007). 'Two-wheeled welding mobile robot for tracking a smooth curved welding path using adaptive sliding-mode control technique. *International Journal of Control, Automation and Systems*, *5*(3), 283–294.

Durairaj, R., & Shanker, J. (2012). *Nano robots in bio medical application*. ieeexplore.ieee.org. Available at: https://ieeexplore.ieee.org/abstract/document/6216236/ (Accessed 24 July 2020).

Epson. (2020). *Epson SCARA LS10-B702S (standard)*. Epson. Available at: https://www.epson.eu/products/robot/epson-scara-ls10-b702s-standard (Accessed 14 August 2020).

Fijany, A., & Bejczy, A. (1993). *Special purpose parallel computer architecture for real-time control and simulation in robotic applications*. Google Patents. Available at: https://patents.google.com/patent/US5218709A/en (Accessed 24 July 2020).

gagileo.org. (2020). *Robotics design process*. Available at: https://www.galileo.org/robotics/design.html (Accessed 24 July 2020).

GeeksforGeeks. (2019). *Fuzzy logic | Introduction*. GeeksforGeeks. Available at: https://www.geeksforgeeks.org/fuzzy-logic-introduction/ (Accessed 14 August 2020).

Hanumara, N. C. (2012). *Efficient design of precision medical robotics (PhD thesis)*. Massachussets Institute of Technology.

Healthcare Administration Degree Programs. (2020). *How are robots changing healthcare?*. Healthcare Administration Degree Programs. Available at: https://www.healthcare-administration-degree.net/faq/how-are-robots-changing-healthcare/ (Accessed 24 July 2020).

Hong, T. S., Nakhaeinia, D., & Karasfi, B. (2012). *Application of fuzzy logic in mobile robot navigation*. Intech Open.

Jayaweera, N., & Webb, P. (2010). Metrology-assisted robotic processing of aerospace applications. *International Journal of Computer Integrated Manufacturing*, *23*(01), 283–296.

Kah, P., Shrestha, M., & Hiltunen, E. (2015). *Vertically articulated (revolute configuration) robot with five.. | Download Scientific Diagram*. Available at: https://www.researchgate.net/figure/ertically-articulated-revolute-configuration-robot-with-five-revolute-joints-Ross-et_fig 3_282540974 (Accessed 14 August 2020).

Kaur, K. (2013a). *Basic robotics—Power source for robots*. Available at: https://www.azorobotics.com/Article.aspx?ArticleID=139 (Accessed 24 July 2020).

Kaur, K. (2013b). *Building an end-effector for robotic control*. Available at: https://www.azorobotics.com/Article.aspx?ArticleID=88 (Accessed 24 July 2020).

Khan, F. (2016). *Internet of things explained. An extremely brief explanation of how IoT works*. Banking, Payments & Fintech. Available at: https://blog.faisalkhan.com/internet-of-things-explained-1d00e022f668 (Accessed 14 August 2020).

Khatib, O. (2019). The age of human-robot collaboration. In *CISM international centre for mechanical sciences, courses and lectures* (p. 9). Springer International Publishing. https://doi.org/10.1007/978-3-319-78963-7_2.

Kiesler, S., et al. (2008). Anthropomorphic interactions with a robot and robot-like agent. *Social Cognition*. Available at: http://anthropomorphism.org/psychology2.html (Accessed 24 July 2020).

Kim, H. S., & Song, J. B. (2014). Multi-DOF counterbalance mechanism for a service robot arm. *IEEE/ASME Transactions on Mechatronics*, *19*(01), 1756–1763.

Kim, S., & Wensing, P. M. (2017). Design of dynamic legged robots. *Foundations and Trends in Robotics*, *5*(2), 117–190. https://doi.org/10.1561/2300000044. Now Publishers.

Lavallee, S., et al. (1992). *Image guided operating robot: A clinical application in stereotactic neurosurgery*. computer.org. Available at: https://www.computer.org/csdl/proceedings-article/robot/1992/00220223/12OmNyTfg62 (Accessed 24 July 2020).

Machine Design. (2020). *The difference between cartesian, six-axis, and SCARA robots*. Machine Design. Available at: https://www.machinedesign.com/mechanical-motion-systems/article/21831692/the-difference-between-cartesian-sixaxis-and-scara-robots (Accessed 14 August 2020).

McClelland, C. (2020). *What is IoT?—A simple explanation of the internet of things*. Available at: https://www.iotforall.com/what-is-iot-simple-explanation/ (Accessed 14 August 2020).

Medical Design Briefs. (2013). *Integrating force sensors into robotic surgery*. Medical Design Briefs. Available at: https://www.medicaldesignbriefs.com/component/content/article/mdb/tech-briefs/16997 (Accessed 24 July 2020).

MIT. (2015). *Robot origami: Robot self-folds, walks, and completes tasks*. YouTube. Available at: https://www.youtube.com/watch?v=ZVYz7g-qLjs (Accessed 14 August 2020).

Murphy, R. R. (2020). *Robots-are-playing-many-roles-in-the-coronavirus-crisis-and-offering-lessons-for-future-disasters*. (22 April 2020). Available at: https://theconversation.com/robots-are-playing-many-roles-in-the-coronavirus-crisis-and-offering-lessons-for-future-disasters-135527.

Nunuparov, A., et al. (2019). Vibration-driven capsubot with an opposing spring: An experimental study. In *CISM international centre for mechanical sciences, courses and lectures* (pp. 126–133). Springer International Publishing. https://doi.org/10.1007/978-3-319-78963-7_17.

Owen-Hill, A. (2019). *5 ways robotics are used in medicine and healthcare*. Available at: https://blog.robotiq.com/5-ways-cobots-are-used-in-medicine-and-healthcare (Accessed 24 July 2020).

Pande, R., et al. (2003). The telecommunication revolution in the medical field: Present applications and future perspective. *Current Surgery*. Elsevier. Available at: https://www.sciencedirect.com/science/article/pii/S0149794403001788 (Accessed 24 July 2020).

Rodríguez León, J. F., et al. (2019). Experiences and design of a cable-driven assisting device for arm motion. In *CISM international centre for mechanical sciences, courses and lectures* (pp. 94–101). Springer International Publishing. https://doi.org/10.1007/978-3-319-78963-7_13.

Rouse, M. (2020). *What is robot?—Definition from WhatIs.com*. Available at: https://searchenterpriseai.techtarget.com/definition/robot (Accessed 24 July 2020).

RS. (2020). *The complete guide to robotic parts*. RS Components. Available at: https://uk.rs-online.com/web/generalDisplay.html?id=ideas-and-advice/robotic-parts-guide (Accessed 24 July 2020).

Russo, M., et al. (2019). An experimental characterization of a parallel mechanism for robotic legs. In *CISM international centre for mechanical sciences, courses and lectures* (pp. 18–25). Springer International Publishing. https://doi.org/10.1007/978-3-319-78963-7_4.

Samiksha, S. (2020). *Robots: 5 Important components of robots*. Available at: https://www.yourarticlelibrary.com/robots/robots-5-important-components-of-robots/5692 (Accessed 24 July 2020).

Sitemap. (2007). *Cylindrical robots*. Available at: https://www.industrial-electronics.com/industrial-electricity-com/4_Cylindrical_Robots.html (Accessed 14 August 2020).

Sivasankar, M., & Durairaj, R. (2012). Brief review on nano robots in bio medical applications. *Advances in Robotics and Automation*. academia.edu. Available at: https://www.academia.edu/download/23665332/ara-1-101.pdf (Accessed 24 July 2020).

Soleimani, F., et al. (2011). *Izobraževanje/education robots and medicine—Shaping and defining the future of surgery, endovascular surgery*. Slovenian Medical Association.

Song, W., & Lee, H. (1999). KARES: Intelligent wheelchair-mounted robotic arm system using vision and force sensor. *Robotics and Autonomous Systems*. Elsevier. Available at: https://www.sciencedirect.com/science/article/pii/S0921889099000317 (Accessed 24 July 2020).

Tahboub, K. (1999). *A compliant semi-autonomous reactive control architecture applied to robotic holonomic wheelchairs*. ieeexplore.ieee.org. Available at: https://ieeexplore.ieee.org/abstract/document/803247/ (Accessed 24 July 2020).

Tahboub, K., et al. (1999). *A semi-autonomous control architecture applied to robotic wheelchairs*. ieeexplore.ieee.org. https://doi.org/10.1109/IROS.1999.812795.

Tavakoli, M., Carriere, J., & Torabi, A. (2020). *Robotics for COVID-19: How can robots help health care in the fight against robotics for COVID-19 update June 5, 2020 how can we help right now? Telehealth to assist frontline healthcare workers*. (March).

Taylor, R. H., & Stoianovici, D. (2003). Medical robotics in computer-integrated surgery. *IEEE Transactions on Robotics and Automation*, *19*(5). https://doi.org/10.1109/TRA.2003.817058.

The Medical Futurist. (2020). *Benefits of robotics in healthcare: Tasks medical robots will undertake*. Available at: https://medicalfuturist.com/robotics-healthcare/ (Accessed 24 July 2020).

Tiwari, P., et al. (2010). *Some non-technology implications for wider application of robots to assist older people*. academia.edu. Available at: http://www.academia.edu/download/45470167/Some_Non-technology_Implications_for_Wid20160509-9727-9pammh.pdf (Accessed 24 July 2020).

Tsui, K. M., & Yanco, H. A. (2007). *Assistive, rehabilitation, and surgical robots from the perspective of medical and healthcare professionals*. aaai.org. Available at: https://www.aaai.org/Papers/Workshops/2007/WS-07-07/WS07-07-007.pdf (Accessed 24 July 2020).

WHO. (2020). *Infection prevention and control during health care when COVID-19 is suspected*. 2019(April). Available at: https://www.who.int/publications-detail/infection-prevention-and-control-during-health-care-when-novel-coronavirus-(ncov)infection-is-suspected-20200125.

Wolf, A., et al. (2005). Design and control of a mobile hyper-redundant urban search and rescue robot. *Advanced Robotics*. Available at: https://www.tandfonline.com/doi/abs/10.1163/1568553053583652 (Accessed 24 July 2020).

Wright, B. (2020). *What is the difference between alkaline and lithium batteries?*. ProProfs Discuss. Available at: https://www.proprofs.com/discuss/q/1709350/what-is-the-difference-between-alkaline-and (Accessed 14 August 2020).

Yoshikawa, T. (1985). *Manipulability of robotic mechanisms*. journals.sagepub.com. Available at: https://journals.sagepub.com/doi/abs/10.1177/027836498500400201 (Accessed 24 July 2020).

Zielińska, T., & Wang, J. (2019). Two methods of EMG analysis for the purpose of exoskeletons and robotic rehabilitation devices. In *CISM international centre for mechanical sciences, courses and lectures* (pp. 110–117). Springer International Publishing. https://doi.org/10.1007/978-3-319-78963-7_15.

CHAPTER 3

Robotic manipulator components' description: Materials and methods

3.1 Introduction

This chapter serves to layout the system of approaches utilized by the researcher to solve the problem stated before and fulfill the aims and objectives. The drawings of the robotic manipulator will also be outlined and explained in this chapter, including the robotic manipulator's specification.

3.2 Materials and methods

3.2.1 Development process for the intelligent robotic machine

The following procedure will be used by the researcher to achieve the aims, objectives, and to come up with the final solution.

3.2.1.1 Concept generation and development

In conceptual design, several design alternatives will be considered and the best alternative will be selected. To better understand the goals of the design, additional activities were included such as reviewing similar products, industry reports, literature, and patents.

3.2.1.2 Concept selection

Each of the solutions will be compared against the criteria deemed important to the problem (such as safety, reliability, ease of use, environment conditions, survivability, and vulnerability). Weights will be allocated to each criterion—this will be done based on consensus. A matrix table will be used to compare the criteria and the first rank (with the highest score) concept will be selected.

Healthcare Systems Design of Intelligent Testing Centers
https://doi.org/10.1016/B978-0-323-99443-9.00007-3

3.2.1.3 *Detailed design of the selected concept*

The first ranked concept will further be developed to meet the design requirements. Scientific and mathematical models such as Gruebler equation, equation of continuity, Lagrange, and Von Mises stress analysis will be used for the development of the selected concept.

3.2.2 Tools for design and prototype testing

See Table 3.1.

3.2.3 Project management

The main project activities have been summarized in Table 3.2.

Table 3.1 Engineering modeling, simulation software, and programming software.

Application		Description
Engineering Modeling and Simulation Software		
	SolidWorks	Like Inventor, this is computer-aided design software, which assists with making 2D or 3D strong models with no intricacy in the financially savvy way. The scientist will utilize this product for costing and maintainability investigations, and dynamic reproduction of detailed, contrasting outcomes and those from similar software like Inventor.
	MATLAB	MATLAB is an elite language for specialized processing. It coordinates calculation, representation, and programming in a simple to-utilize environment where issues and arrangements are communicated in natural numerical documentation. The scientist will utilize it for algorithm improvement, modeling, validation, and simulation.
Programming software		
	Python	The researcher will program the robot framework in Python. Python is a deciphered, object-situated, significant level programming language with dynamic semantics, undeniable level inherent information structures, joined with dynamic composing and dynamic restricting.

Table 3.2 Project time line.

	Description of work	Start and end dates
Phase One	Investigation and identification of the problem	15/07/2020: 01/07/2021
Phase Two	Concept generation of robotic system components	10/09/2020: 05/07/2021
Phase Three	Detailed design of the selected concept and prototype testing	5/10/2020: 01/07/2021

3.2.4 Project time frame

See Table 3.2.

3.2.5 Deliverables

The deliverables include a well-written project report and the simulation results which will be drafted and represented to justify design parameters, component selection, and the prototype. The feasibility of the project will be shown on the economic report.

3.3 Possible solutions

Three concepts have been generated in this research and they are outlined as follows. All the concepts were developed to meet the aims and objectives of this research. The general principle of operation together with advantages and disadvantages of each concept will also be discussed.

3.3.1 Concept one

This first robotic design was initially designed based on using the least number of materials possible to achieve the aims and objectives. This design is effective while economical and can be manufactured and assembled at an affordable cost. The robotic machine consists of a single arm that will be attached on top of a table surface which in turn will be placed at the hospital entrance. The basic robotic arm designed with four degrees of freedom seemed ideal because of its simplicity. The robotic arm has an ability to rotate about its base and can collect a swab placed in the table, using the gripper and use it to collect the specimens from a patient's nostril or throat. The patient being tested will be seating on a chair next to the table. After collecting the swab, the arm will place it in a container. Thereafter, the arm will mix the swab with chemicals and start the testing process. The robotic arm will do the antigen test, where it will use a dropper to drop a mixture of a chemical and sample in the slide plate. Through the use of sensors, the system will determine whether someone is positive or negative and the results will be sent to one's email address. The design from attachment to the end effector of the arm consists of four joints and four links (which are frames), with three joints attached with servo motors to control the arm's movement (see Fig. 3.1). The robot system will be controlled by the raspberry pi.

Advantages

✓ Relatively economical design.

✓ Easy to manufacture.

Disadvantages

✓ The design is not safe due to the frame links which might injure a patient during testing.

✓ The electrical components are not enclosed and might cause short circuiting while testing.

✓ The testing process is relatively slow as the single arm has to perform everything.

✓ Contamination of the testing equipment as there is not barrier between the arm and the patient.

✓ Lack of flexibility due to small number of degrees of freedom.

Fig. 3.1 Concept one isometric view. *(Credit: Marvellous Moyo and Tawanda Mushiri.)*

3.3.2 Concept two

This next design consists of similar materials and some parts of the first design but with different arm design. Also, in this design, the robotic arm control is the same as the first design. The unique feature in this design is a dual robotic arm system, with the links of the arm being almost cylindrical in shape to enhance safety of the system. These two robotic arms are attached at two opposite ends of the table and will help each other out to carry out the testing process. One arm will collect the swab from patients sitting next to the table and put them in the container while the other will be mixing sample with chemicals and doing the testing. This design was meant to improve the testing rate of the first design. Therefore, in addition to the antigen test done by the first design, since there are two arms now then if someone test negative for the antigen test the RT-PCR test will also be done for the same sample. The arm which is doing the testing will take the sample with chemicals and place it in PCR machine for testing and the results will be sent via email address. The two robotic arms have each five degrees of freedom to increase the flexibility when testing (see Fig. 3.2).

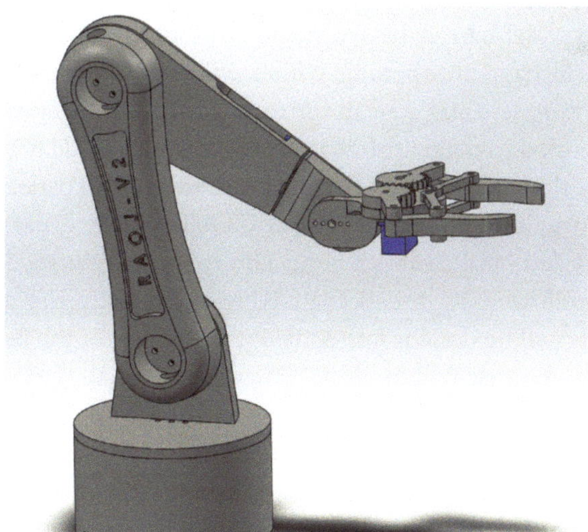

Fig. 3.2 Concept two isometric view. *(Credit: Marvellous Moyo and Tawanda Mushiri.)*

Advantages

- ✓ Improved safety due to enclosure of electrical components.
- ✓ Faster testing rate due to dual arm configuration.
- ✓ The system is reliable since it can perform two tests for a single sample.
- ✓ Improved flexibility because of increased number of degrees of freedom.

Disadvantages

- ✓ The design is relatively expensive due to additional components.
- ✓ Contamination of the testing equipment as there is not barrier between the arm and the patient.
- ✓ The design is relatively less efficient due to other links which are almost rectangular in shape increasing drag during operation.

3.3.3 Concept three

The third design was designed based on making the most reliable robotic machine and the safest testing environment while keeping all the objectives in mind. Also, in this design, the robotic arm control is the same as the previously mentioned designs. This design like the second concept consists of two robotic arms but there is a barrier between the arms and the patient. Therefore the air between the patient and the two robotic arms will not be able to mix. The barrier between the arms and the patient sitting next to the table will have an opening which will open when one arm wants to collect the sample using a swab for testing. During this process, a compressor will be used to make sure that air inside the arm side has higher pressure so that the flow of air will be from robotic arm side to patient's side. This robotic system will do all three types of COVID-19 test, namely: antigen test, RT-PCR test, and antibody test. The robotic system will first do the antigen test, followed by the RT-PCR test and lastly the antibody test and all the tests will be done if someone tests negative for the preceding test, otherwise only the first will be done. Therefore, initially, when the arm is collecting samples it will collect enough samples for both antigen and RT-PCR test and a blood sample for the antibody test. The two robotic arms have each six degrees of freedom to increase the flexibility when testing. The shapes of the links have been made more cylindrical to improve efficiency and safety (see Fig. 3.3).

Advantages

- ✓ Improved efficiency due to cylindrical links.
- ✓ Highly reliable and informative since the system can perform three tests.

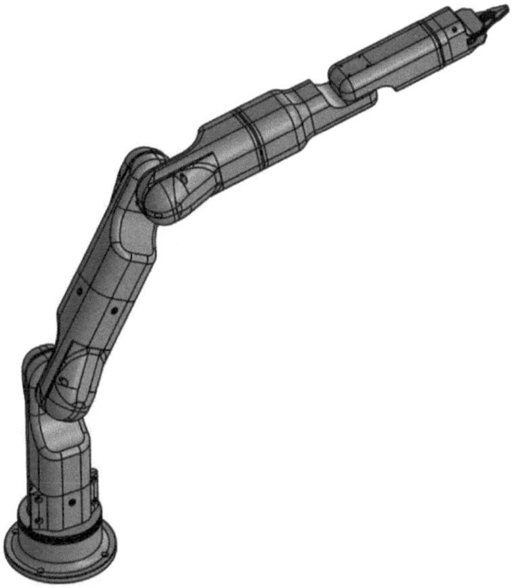

Fig. 3.3 Concept three isometric view. *(Credit: Marvellous Moyo and Tawanda Mushiri.)*

✓ Improved safety due to enclosure of electrical components.
✓ Faster testing rate due to dual arm configuration.
✓ Improved flexibility because of increased degrees of freedom.
✓ Equipment contamination is not possible as there is a barrier between the robots and patients.

Disadvantages

✓ The design is relatively expensive due to additional components.
✓ The system is very complex.

3.4 Concept selection

The choice of the final solution depends upon some factors which affect the design's ability to meet the objectives of the project in the best way. The criteria which will be used in the selection of the best concept are listed as follows:

A. Safety
B. Reliability
C. Functionality
D. Efficiency
E. Weight
F. Cost

G. Ease of Manufacture
H. Ease of Maintenance
I. Simplicity of Design
J. Life Span

In order to select the best solution, different concepts will be compared according to how much they meet the earlier criteria. The first step will be the ranking of the design criteria in order of their relative significance. This will be done using the binary dominance method (see Table 3.3). This is done by listing the criteria on the vertical and horizontal axes of a table. After comparison of a pair of criteria, a 0 or 1 (binary) will be placed in the boxes of the matrix based on the relative significance.

Table 3.4 presents the selection criteria in order of their importance. It shows what is to be considered first in making the final design selection and a new order for the criteria.

In the concept weighting stage, the total number of a criterion will be taken as its weight. Each concept will then be rated, on a 1–10 scale, for each criterion. The final weight which will be allocated to a concept under a specific criterion will be the product of the respective rating and weight of criterion (see Table 3.5).

From Table 3.5, the third concept ranked number one and shall now be developed according to the methodology described before.

Table 3.3 Binary dominance matrix.

	A	B	C	D	E	F	G	H	I	J	Total
A. Safety	–	1	1	1	1	1	1	1	1	1	9
B. Reliability	0	–	1	1	1	1	1	1	1	1	8
C. Functionality	0	0	–	0	1	1	1	1	1	1	6
D. Efficiency	0	0	1	–	1	1	1	1	0	1	6
E. Weight	0	0	0	0	–	1	1	0	1	0	3
F. Cost	0	0	0	0	0	–	1	0	1	1	3
G. Ease of manufacture	0	0	0	0	0	0	–	0	1	1	2
H. Ease of maintenance	0	0	0	0	1	1	1	–	0	0	3
I. Simplicity of design	0	0	0	1	0	0	0	1	–	1	3
J. Life span	0	0	0	0	1	0	0	1	0	–	2

Table 3.4 Criteria ranking and revised order in terms of their importance.

Criteria	Total	Ranking	New order in terms of importance
A. Safety	9	1	A. Safety
B. Reliability	8	2	B. Reliability
C. Functionality	6	4	C. Functionality
D. Efficiency	6	3	D. Efficiency
E. Weight	3	5	E. Weight
F. Cost	3	6	F. Cost
G. Ease of manufacture	2	9	G. Simplicity of design
H. Ease of maintenance	3	7	H. Ease of maintenance
I. Simplicity of design	3	8	I. Ease of manufacture
J. Life span	2	10	J. Life span

Table 3.5 Concept weighting and scoring.

		Concept rating			Concept weight		
Criteria	Weight	One	Two	Three	One	Two	Three
A	9	5	6	9	45	54	81
B	8	7	8	9	56	64	72
C	6	5	7	9	30	42	54
D	6	6	6	9	36	36	54
E	3	8	7	6	24	21	18
F	3	8	6	5	24	18	15
G	3	7	6	5	18	18	15
H	3	7	6	6	21	18	18
I	2	8	7	6	16	14	12
J	2	5	6	8	10	12	16
Total					***280***	***297***	***355***
Rank					***3***	***2***	***1***

The underlined number is the chosen value.

3.5 Drawing description

Fig. 3.4 shows the exploded view of the robotic arm showing exactly how the links are connected to each other. The robotic arm consists of six links plus an end effector and these include: base link **(1)**, link 1 **(2)**, link 2 **(3)**, link 3 **(4)**, link 4 **(5)**, link 5 **(6)**, and end effector **(7)**. Fig. 3.5 shows all of the components of the robotic arm.

These components include base (26), link 1 part one (24), link 1 part two (22), link 1 side cover (25), servo cover (28), big servo (21), small servo (12),

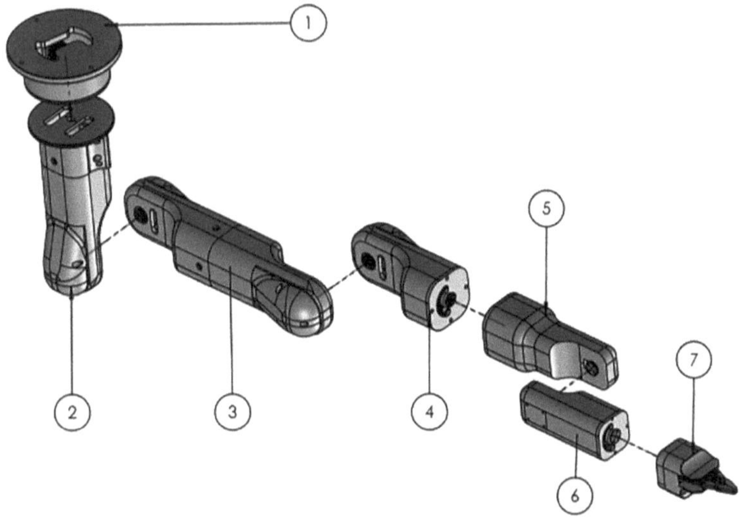

Fig. 3.4 Robotic arm exploded view showing each link connection. *(Credit: Marvellous Moyo and Tawanda Mushiri.)*

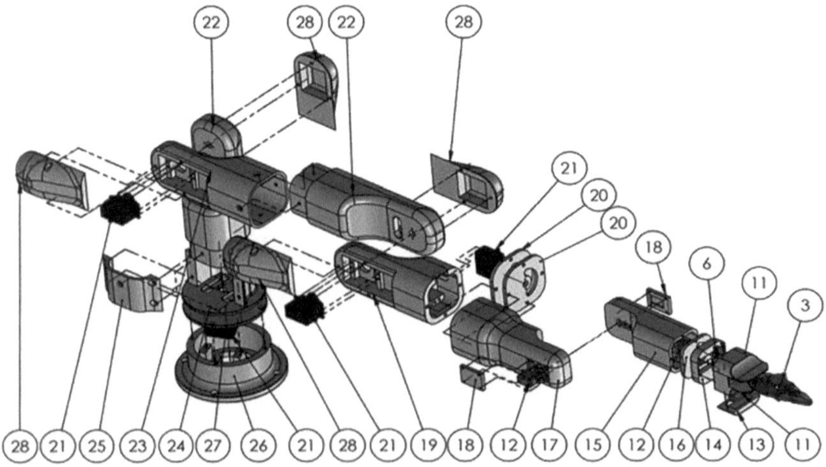

Fig. 3.5 Robotic arm exploded view showing the connection of each component. *(Credit: Marvellous Moyo and Tawanda Mushiri.)*

link 2 (23), link 3 (19), link 3 and 4 cover (20), link 4 (17), small servo side cover (18), link 5 (15), link 5 cover (16), end effector cover (11), and gripper (3). The big servo motors are used to actuate the first links from the base and the smaller servo is used to actuate the spherical wrist and gripper. By using of the earlier component connections, the arm is able to

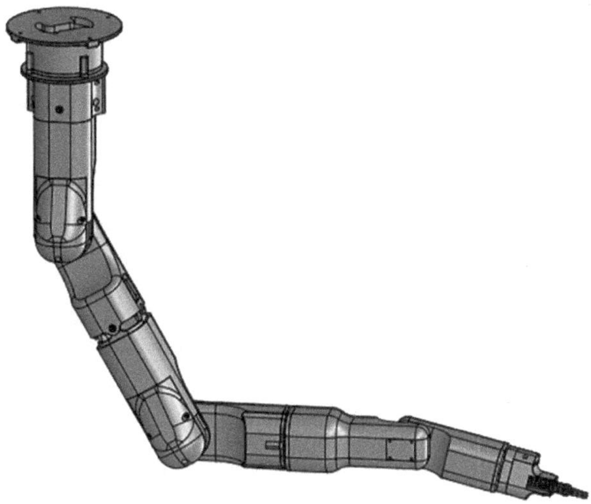

Fig. 3.6 Isometric view of the optimized robotic manipulator. *(Credit: Marvellous Moyo and Tawanda Mushiri.)*

maneuver inside the booth and collect a sample from the patients' nose and then do testing. Fig. 3.6 shows the same robotic manipulator with optimized components link, also see Appendix C for the detailed engineering drawings. The detailed optimization of the manipulator will be covered in later chapters.

3.6 Technical specifications

The technical specification of the robotic manipulator that will be used to test COVID-19 virus is summarized in Table 3.6.

Table 3.6 Technical specifications.

Specification	Description
Robotic arm and actuation	
Length	1.25 m
Weight (including end effector)	4.2 kg
Number and types of joints	6 revolute joints
Types of actuators	MG996R and SG90 Micro servo motors
Power rating	1400 W

Continued

Table 3.6 Technical specifications—cont'd

Specification	Description
Number of links	7
DOF	6
Power source	Battery: lithium-polymer batteries (LiPos)
Types of sensors	▪ *Robotic arm sensors*: position sensor, velocity sensors, vision sensors such as cameras (mono or stereo), force sensors, proximity sensors, switches. ▪ *Medical sensors:* temperature sensors, pulse sensors—optical heart rate monitors (LED), respiratory sensor. ▪ *Obstacle avoidance sensor:* ultrasonic sensor.
Microcontroller	Raspberry pie
Working principle	Fully autonomous
Wireless communication	Wi-Fi modules
Material	Kevlar epoxy (53%)
End effector	
Type	Gripper

3.7 Conclusion

This chapter outlined the step-by-step design process that was carried out to come up with an intelligent robotic machine for testing COVID-19. It outlines the procedures used to gather data and how this data is used to achieve the aims and objectives. In addition, three concepts were generated and concept number three was selected for further development. The general components of the system were also discussed in this chapter and lastly the specifications of the medical robotic manipulator were specified.

CHAPTER 4

COVID-19 effects and current testing techniques used in hospitals: Healthcare workers' perspectives

4.1 Introduction and background

The coronavirus (COVID-19) pandemic is at present a worldwide health emergency; in excess of 179 million cases had been affirmed worldwide and in excess of 3.87 million deaths had been reported at the time of composing this report (WHO, 2021). COVID-19 virus, a disease brought about by serious deadly respiratory phenomenon COVID 2 (SARS-CoV-2), is exceptionally infectious and has spread quickly around the world. In serious cases, COVID-19 virus can prompt intense respiratory distress, numerous organ failures, and in the end death. Nations are hustling to moderate the propagation of the infection by diagnosing and treating patients in the beginning phase, just as isolating individuals who are at high danger of exposure because of close contact with patients with affirmed infection (Khyber et al., 2020; Watkins-Hayes, 2019). Moreover, early determination and rapid treatment are significant to saving the existence of patients with affirmed infection.

The COVID-19 pandemic is a remarkable medical care emergency that has required medical and research facility experts to quickly adjust to new data, improve, and, in certain circumstances, execute practices that would not be considered under ordinary conditions (Garriga et al., 2020; Hollander & Carr, 2020; Shaker et al., 2020; Vieta, Pérez, & Arango, 2020). The rise of SARS-CoV-2 has constrained clinical research facilities and test manufacturers to develop new diagnostic assays in a time span that already would not have been considered practical. As testing choices have opened up, medical care experts and diagnostics specialists have needed to figure out how to best apply these devices to determine and oversee patients who have COVID-19 virus and moderate the spread of infection (Israfil et al., 2021;

Healthcare Systems Design of Intelligent Testing Centers
https://doi.org/10.1016/B978-0-323-99443-9.00005-X

Kim, 2020; Zoghbi et al., 2020). A portion of these exercises have shown routine testing ways to be effective, while others have instructed, saying that another pathogen, for example, SARS-CoV-2, may require another diagnostic playbook.

As SARS-CoV-2 has spread worldwide, it has required lab experts to wander into uncharted area. Not exclusively is testing being utilized to determine patients who have symptoms, it is being consolidated into local, state, and government techniques to resume the economy (Jang, Han, & Rhee, 2020; van Doremalen et al., 2020). Organizations, schools, and surprisingly athletic groups are thinking about how research facility testing can be utilized to show that asymptomatic employees, clients, understudies, and athletics are safe to get back to typical exercises. A few medical clinics and centers are using SARS-CoV-2 testing before specific techniques and medical procedures to lessen nosocomial transmission and forestall poor results in high-risk patients (Chakraborty, Jamal, & Battineni, 2020). This pattern toward enormous scope screening of asymptomatic people has put a mind-blowing trouble on the worldwide testing framework and has created difficulties with respect to how testing ought to be used and results interpreted. Accordingly, it is anything but amazing that various demonstrative vulnerabilities and debates have emerged.

COVID-19 virus is regularly affirmed by viral nucleic acid recognition utilizing reverse transcription-polymerase chain reaction (RT-PCR) (Prabhakar & Lakhanpal, 2020; Sen & Rout, 2021). Nevertheless, the sensibility of RT-PCR may not be adequately high; it goes from 37% to 71% as per early reports (Ray & Subramanian, 2020). In this manner, RT-PCR can bear the cost of a generous number of false negative outcomes because of insufficient specimen collection, wrong extraction of nucleic acid from the specimen, or assortment at a too soon phase of infection.

4.2 COVID-19: The global health emergency

The World Health Organization (WHO) has termed the novel coronavirus COVID-19 as a worldwide pandemic which is a very dangerous threat to human life (Sohrabi et al., 2020). The origins of this virus were traced to the supermarket in Wuhan, China, in December 2019. Currently the COVID-19 virus has no approved specific treatment and there is no treatment for the infection. However, treatments and vaccines are being investigated on as we speak. In the meantime, the focus is on managing symptoms as the virus runs its course. COVID-19 has some serious complications such as the new type of pneumonia called 2019 novel coronavirus-infected pneumonia (NCIP). Researchers have shown that people who have developed COVID-19 have the following complications:

- ✓ acute respiratory distress syndrome (ARDS)
- ✓ irregular heart rate (arrhythmia)
- ✓ cardiovascular shock
- ✓ severe muscle pain (myalgia)
- ✓ fatigue or exhaustion
- ✓ heart attack or damage

4.2.1 Symptoms of COVID-19

Doctors are researching on this virus every day and so far, some results have shown that COVID-19 does not initially show any symptoms for some individuals, therefore testing people is essential to know who really has the virus (Lan et al., 2020b). The incubation period of the novel coronavirus is 2 days up to 2 weeks. The usual and less likely symptoms of COVID-19 are summarized in Table 4.1 (Healthline, 2020; Hollander & Carr, 2020).

These symptoms are subjective, therefore, maybe more severe in some people. One should call emergency medical services if or someone develops the following symptoms.

- ✓ difficulties breathing
- ✓ blue lips or face
- ✓ continuous pain or pressure in the chest
- ✓ confusion
- ✓ too much drowsiness

Table 4.1 Symptoms of COVID-19 virus.

Common symptoms	Less common symptoms
Shortness of breath	Chills
A cough that gets more severe over time	Repeated shaking with chills
A low-grade fever that gradually increases in temperature	Loss of taste
Fatigue	Muscle aches and pains
	Headache
	Sore throat

4.2.2 COVID-19 causes and prevention

The novel coronavirus or any coronavirus has been discovered to be zoonotic. Generally, what this term implies is that they initially develop in creatures before being transmitted to humans. This virus can be transmitted from animals to humans when a human comes in close contact with the infected person. After that, the coronavirus can be transmitted from human to human through respiratory fluids. These droplets are released when one is talking, coughing, or sneezing. These droplets contain the virus inside them and can be inhaled into the respiratory tract and there after leading to infection. Therefore someone is at high risk of contracting the COVID-19 virus when one comes into contact with droplets from an infected person. COVID-19 can be checked similarly to other diseases caused by viral infections, that is, utilizing a blood, saliva, or tissue sample. However, a lot of tests utilize a cotton swab to collect a sample from the inside of one's nostrils. Therefore, from the foregoing statements, the most efficient way to deter the spread of coronavirus is to avoid or reduce contact with individuals who are showing symptoms of COVID-19 or any respiratory infection (Gandhi, Yokoe, & Havlir, 2020a; Lan et al., 2020a; Reynolds & Adhikari, 2020; Wang, Ng, & Jama, 2020). In addition, it is crucial to practice good hygiene all the time. This suffice to say, there is a desperate need to test more people to know who should be quarantined from others or whom should one isolate themselves from, thereby mitigating the spread of the virus (Beeching, Fletcher, & Beadsworth, 2020; Bhatraju & Ghassemieh, 2020; Gandhi, Yokoe, & Havlir, 2020b). Fig. 4.1 shows ways to adapt to the COVID-19 pandemic.

4.3 COVID-19 testing techniques

There are two different types of test used to test COVID-19 and are as follows (Avert, 2020; FDA, 2020):

✓ diagnostic tests and

✓ antibody tests.

A diagnostic test is used to show if one has the infection. So far, there are two sorts of tests used to detect the virus, the molecular tests used to detect the virus genetic material such as the RT-PCR, and antigen tests that detect unique proteins on the virus surface. An antibody test detects antibodies generated by the body's immune framework in response to a threat, such as unique virus.

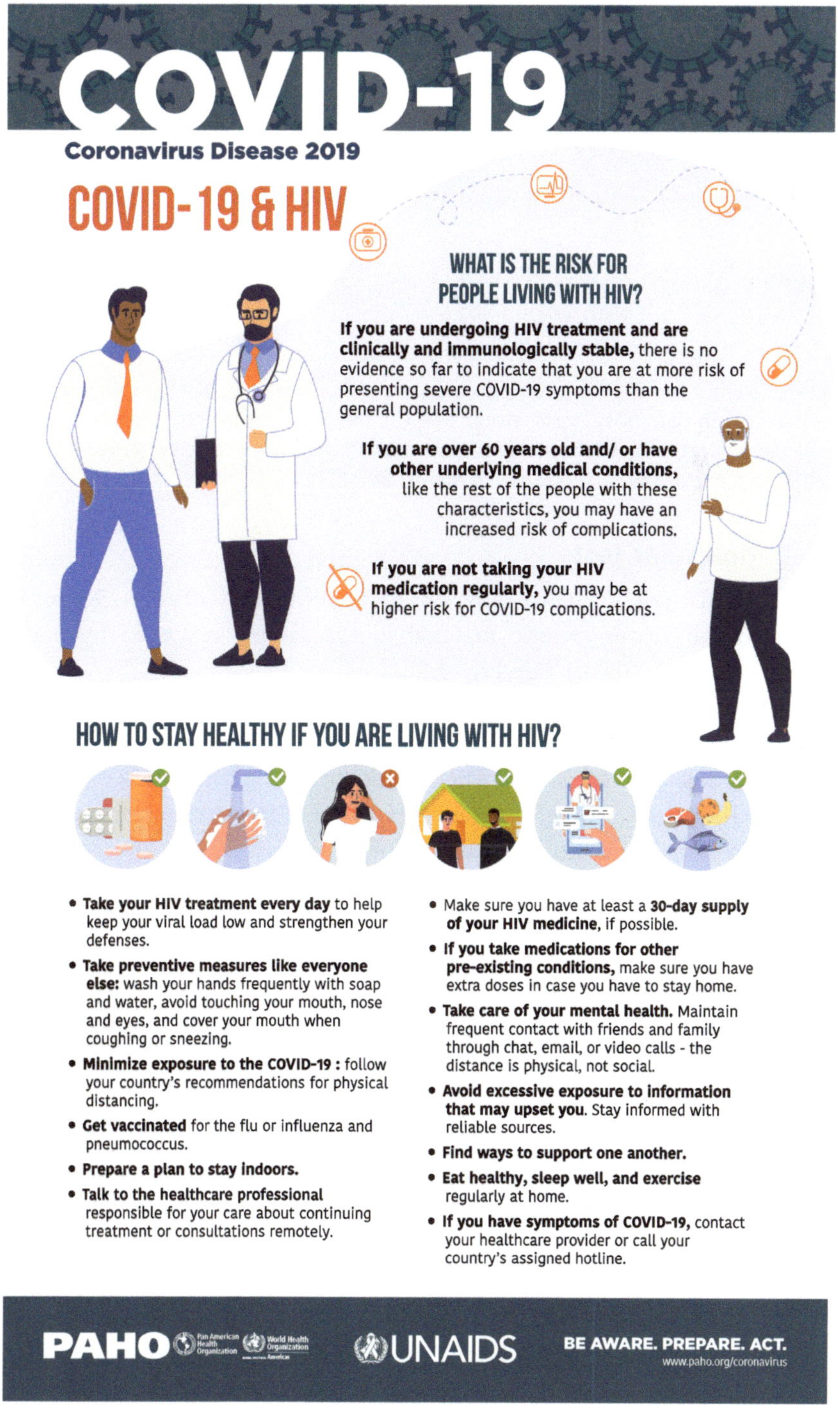

Fig. 4.1 COVID-19 virus prevention: Treating COVID-19 like HIV. *(Courtesy Avert. (2020).* Infographics on coronavirus (COVID-19). *Avert. Available at: https://www.avert.org/coronavirus/infographics. (Accessed 14 August 2020).)*

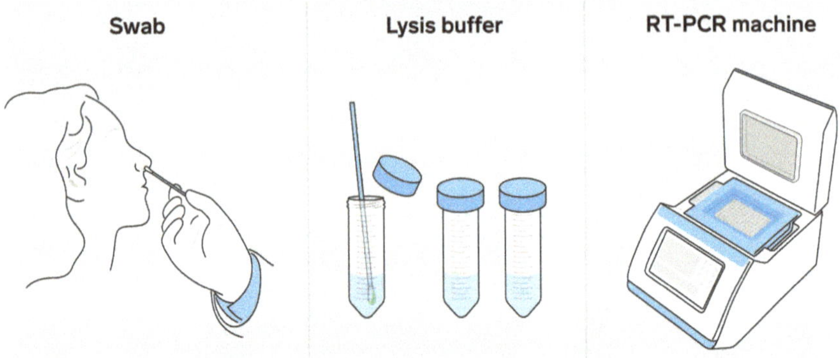

Fig. 4.2 Layout for RT-PCR testing process. *(Courtesy Brueck, H., & Lee, S. (2020).* How coronavirus throat, nose tests work: RT-PCR method, explained. *Business Insider. Available at: https://www.businessinsider.com/how-coronavirus-throat-tests-work-rt-pcr-method-explained-2020-4. (Accessed 14 August 2020).)*

4.3.1 Molecular tests

The molecular test is used by most labs to diagnose the COVID-19 virus in regard to the virus's genetic material in the sample retrieved from the patient's nostril or throat (see Fig. 4.2).

The current basic steps in molecular testing are as follows (Ellis, 2020; FDA, 2020):

1. Firstly, health professionals such as doctors or pharmacists order the COVID-19 test. Therefore all of the COVID-19 tests require a prescription from medical professionals.
2. A health professional or one self uses a specialized swab to retrieve mucus from one's nose or throat.
3. The swab must then be enclosed inside a sterile container and transported to the lab.
4. To get accurate results, the swab must be kept at a certain temperature range (see Table 1.3) during the shipping process.
5. In the lab, chemicals are mixed with the swab by a technician, to retrieve genetic material of any virus on the swab.
6. Specialized chemicals termed primers and probes and technologically advanced machine are used to conduct various regulated heating cooling processes to convert the virus's RNA to DNA and then make million copies of the DNA.
7. The binding of this DNA to a specific probe produces a unique type of light that can be detected by the machine and the test reveals a positive result for infection with SARS-CoV-2, the cause of COVID-19 virus.

4.3.1.1 Limitations of the molecular diagnostic test

✓ The inability of the swab to collect the virus from one's nose or throat.

✓ Contamination of the sample during collection and analysis.

✓ The sample might be kept at an incorrect temperature range before analysis.

✓ The failure of the chemicals utilized to retrieve the virus's genetic material and create copies of DNA to work correctly.

4.3.2 Serologic (antibody) testing

Antibody tests provide relatively quick results but it is not utilized to check an active infection. In contrast to PCR tests (molecular test), which swabs to detect the COVID-19, antibody test uses blood samples (see Fig. 4.3) (FDA, 2020; Kent, 2020; Patient Power, 2020). This is due to the fact that small amount of coronavirus circulates through the blood compared to the respiratory tract.

4.3.2.1 Strengths and limitations of antibody tests

✓ A precise antibody test would assist the scientists and researchers gather the data on who has gotten the virus, including those who did not show any symptoms of being sick. In addition, it could also help develop the treatment to treat the virus (Patient Power, 2020).

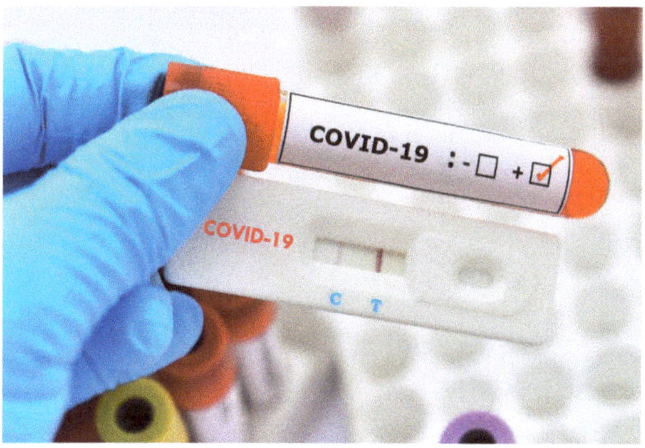

Fig. 4.3 Pictorial view of antibody testing. *(Courtesy Patient Power. (2020).* The pros and cons of the COVID-19 antibody tests. *Patient Power. Available at: https://patientpower.info/navigating-cancer/support-and-resources/the-pros-and-cons-of-the-covid-19-antibody-tests. (Accessed 14 August 2020).)*

✓ The antibody test may yield false negatives if conducted too soon, before the antibodies have developed (Ellis, 2020). The antibody test can also yield false positives, that is, it can detect antibodies which have been produced for other type of coronaviruses such as the SARS-CoV (2003) rather than COVID-19.

4.3.3 Antigen test

Immune system of most vertebrates works based on the principle that proteins that are not encoded by one's genes are maybe from a pathogen and must be destroyed (Wu, 2020). Therefore when a foreign protein is detected, white blood cells (specifically B-cells) create antibodies to trap and destroy the protein. These antibodies are proteins with Y like shape that grabs foreign proteins using their arms. This means foreign proteins trigger this process and it is alluded to as antigen because it is an antibody creator (Kent, 2020) (see Fig. 4.4).

4.3.3.1 Working principle of the antigen test

From the name, the test looks for antigens and antibodies are used to identify them. Antibodies hunt for proteins embedded in the COVID-19's virus surface. If the sample contains a viral protein, it will be detected by the antibodies and the patient is mostly likely infected by the virus (see earlier figure).

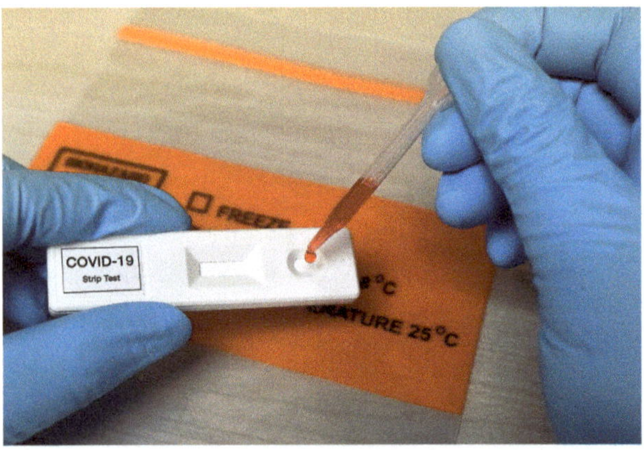

Fig. 4.4 Antigen testing. *(Courtesy WebMD. (2020).* Coronavirus (COVID-19) antibody tests: How it works & how to get one. *Available at: https://www.webmd.com/lung/antibody-testing-covid-19#1. (Accessed 14 August 2020).)*

1. The first step in antigen test is sample collecting like in a molecular test using a swab.
2. The swab is then dipped in a liquid to release the virus.
3. The liquid is then applied to test slide surface coated with antibodies.
4. Antibodies are stuck on the slide and grab any COVID-19 virus that is on the sample.
5. After that, another mixture of antibodies is applied to the slide. The new antibodies have been chemically adapted with a dye such that they can be seen with a naked eye or detectable by fluorescent light.
6. If the sample has viral antigen proteins, these proteins will now be attached with two antibodies, one that attaches viral proteins to the test kit and another to make them visible.
7. The higher the coronavirus antigen, the more visible is the dye, showing that the patient is affected by SARS-CoV-2. If the dye is not detectable, it means the patient is not infected by SARS-CoV-2 or viral protein was not enough in the sample.

4.3.3.2 Strengths and limitations of antigen test

✓ The main benefit of antigen test is that it produces results far much quicker and requires less expertise to perform than reverse transcriptase-polymerase chain reaction (RT-PCR) tests.

✓ Antigen test has relatively low accuracy. This is because the test looks for evidence of virus so it requires a lot of virus protein to produce a detectable result. Therefore if a sample has inadequate viral protein or a patient has low-graded infection, the test might show a false negative result.

4.3.4 Summary of COVID-19 testing techniques

Table 4.2 presents the summary of COVID-19 testing techniques.

4.3.5 Importance of testing COVID-19

Testing is very essential in managing the spread of COVID-19 virus. This is because it helps determine who needs isolation from other people and whom to distance oneself from. Furthermore, countries like South Korea and Germany who had high testing capability seem to have a better course from the pandemic (Ellis, 2020; FDA, 2020). Therefore testing can give a clearer picture of how the virus is spreading, affecting different types of people, and the direction the whole situation of the virus is taking.

Table 4.2 COVID-19 testing techniques (Kent, 2020).

	Molecular test	Antigen test	Antibody test
Another name	Diagnostic test, viral test, molecular test, nucleic acid amplification test (NAAT), RT-PCR test, LAMP test	Rapid diagnostic test (Some molecular tests are also rapid tests.)	Serological test, serology, blood test, serology test
How to retrieve the sample	Nasal or throat swab (most tests) Saliva (a few tests)	Nasal or throat swab	Finger stick or blood draw
Time line to receive results	Same day (some locations) or up to a week	1 h or less	Same day (many locations) or 1–3 days
Is another test needed	This test is typically highly accurate and usually does not need to be repeated.	Positive results are usually highly accurate but negative results may need to be confirmed with a molecular test.	Sometimes a second antibody test is needed for accurate results.
What does it show	Diagnoses active coronavirus infection	Diagnoses active coronavirus infection	Shows if you've been infected by coronavirus in the past
What it cannot achieve	Show if you ever had COVID-19 or were infected with the coronavirus in the past	Definitively rule out active coronavirus infection. Antigen tests are more likely to miss an active coronavirus infection compared to molecular tests. Your healthcare provider may order a molecular test if your antigen test shows a negative result but you have symptoms of COVID-19.	Diagnose active coronavirus infection at the time of the test or show that you do not have COVID-19

4.4 Materials and methods

4.4.1 Research design

A case study has been utilized in this research. The motivation behind the case study is to permit analysts to get and inspect data in a unique situation or phenomenon. The chosen cases can be doctors, biomedical engineers, and staff who are members of the healthcare community (Creswell, 2011). Case study data is utilized to portray a case top to bottom, everything considered, extensively (Radley & Chamberlain, 2012; Yin, 2013). In this investigation, the current COVID-19 testing strategies utilized in Zimbabwean hospitals were investigated inside and out. A literature survey was done to decide the conceptual and operational meanings of the research focus, specifically data relating to COVID-19 testing techniques and effects in hospitals. The development of the instruments for data collection depended on operational definitions, and the instruction was isolated into three fundamental parts that included sociodemographic, probability, and severity; COVID-19 personal experience; COVID-19 testing; COVID-19 effects; medical robots; and debriefing. The collection of data was done in stages, through online surveys to get rich information. Data analysis was done side by side with setup systems for examining observation items and pointers (Singh, Fox, & Brown, 2016; Yin, 2013).

4.4.2 Participants

The encounters and opinions of view of healthcare workers during COVID-19 virus testing were investigated utilizing a questionnaire. A blend of convenience sampling (researchers themselves reached doctors to invite them to take part in the study) and snowball sampling (the participating doctors were requested to forward the questionnaire to their colleagues through social media or WhatsApp groups) was utilized. The goal was to guarantee maximum participation. The populace for the study was chosen in a nonprobabilistic manner and consisted of 50 doctors from Sally Mugabe Central Hospital. Most of the respondents were male, most had less than 5 years of experience (93.3%) while 6.7% had 5–10 years of experience (see Table 4.3). Nevertheless, 20% of the respondents were female and 80% male, 60%, 20%, 13.3%, and 6.7% were undergraduates, postgraduates, doctorates, and professors, respectively. Forty percent were doing or did biomedical engineering while 60% did medicine and surgery.

Table 4.3 Sociodemographic attributes of respondents.

Variable	Category	Percentages
Gender	Male	80%
	Female	20%
Field of study or profession	Biomedical engineering	40%
	Medicine and surgery	60%
Academic level	Professor	6.7%
	Doctorate	13.3%
	Postgraduate	20%
	Undergraduate	60%
Years of experience	Less than 5 years	93.3%
	5–10 years	6.7%
	Above 10 years	0.0%

4.4.3 Instruments

A questionnaire created by the researchers was utilized to investigate the perspectives of Sally Mugabe Central Hospital healthcare workers during the COVID-19 virus testing and challenges the hospital was facing during the COVID-19 pandemic (see Appendix A). The questions in the questionnaire were created dependent on encounters of the researchers and a literature survey (Alexandrovsky et al., 2020; Tasdemir, 2010). It merits referencing the encounters of researchers here: they both work at Sally Mugabe Central Hospital, they are designing simulation hub to help healthcare workers during COVID-19 testing. In addition, they both are doing projects at Parirenyatwa Central Hospital, mainly to do with application of robotics in medicine.

During the creation of the questionnaire, opinions of colleagues in a similar line of work were sought. After the updates, the questionnaire was framed into the last form. The questionnaire comprises of 23 questions, two of which are open ended. The questionnaire included questions designed to solicit healthcare workers' input on the current COVID-19 testing techniques being used at their hospital.

4.4.4 Data collection procedure

The questionnaire was conveyed to the healthcare workers using online platforms and took roughly 5 min to answer. Owing to the pandemic, the data assortment process could not be led face to face; hence, the respondents were reached through online platforms, for example, social media and messaging applications. It is important that the greatest bias in our sample was brought about by the online organization of the questionnaire, as healthcare workers who had internet access participated in the study. The cycle started with the circulation of the data collection devices to doctors who were exposed to the COVID-19 patients. Then, at that point the questionnaire was sent to other healthcare workers using social and online media.

4.4.5 Data analysis

Descriptive statistics were utilized to break down the quantitative data. In the wake of deciphering the percentage of distribution of the responses to the questionnaire, graphics and tables were utilized to make it more justifiable. The qualitative data were investigated utilizing the content analysis technique. The reason for utilizing content analysis was to sort out and separate importance from the information gathered and to make sensible inferences (Berg, 2009). Thematic coding was utilized for the qualitative inquiry. Along these lines, themes and subthemes were determined. The data gathered from the open-ended inquiries were coded by the researchers and afterward comparable thoughts were recombined with the thematical coding approach. In the understanding of qualitative data, respondents' quantitative responses were incorporated, and it was intended to set up a connection among quantitative and qualitative data. Through this, the data was dissected with a holistic point of view.

With an end goal to resolve methodological issues in this investigation, authenticity and transferability of the study were considered. Qualitative researchers look to foster authenticity, which is considered as the essential measure for building up the legitimacy. In this investigation, multiple ways to establish authenticity were set up. The part of researchers in the study, the cycle of creating data collection devices, data assortment, and investigation strategies were clarified exhaustively to guarantee authenticity of the research. The data assortment instrument was created dependent on encounters of the scientists and a literature survey; then expert perspectives were received from associates in various fields. Triangulation procedures (Morse et al., 2002; Pyett, 2003) were carried out by gathering data utilizing numerous strategies for quantitative and qualitative methodologies (counting multiple choice and open-ended questions). Tables and graphics were utilized to build the intelligibility of the quantitative information. The examination of open-ended questions was utilized to support quantitative information. Ultimately, the discoveries were upheld by direct citations of the respondents. To improve transferability of the investigation, detailed description is given (Gao & Xu, 2008; Yin et al., 2009), peer questioning and negative case examination were utilized (Lincoln & Guba, 1985).

4.5 Results

4.5.1 COVID-19 virus testing

64.3% of the healthcare workers confirmed that they have been tested for COVID-19 virus while 35.7% affirmed that they have not yet been tested, as shown in Fig. 4.5

As shown in Fig. 4.6, 46.7% of the healthcare workers at Sally Mugabe Central Hospital stated that any individual can be tested for COVID-19 virus, while 53.3% were against that idea.

According to 12.5% of the healthcare workers, they cannot test anyone because one needs an approval from the doctor first, while 25% suggested that one needs to show signs of symptoms first. Seventy-five percent of the healthcare workers stated that anyone cannot just be tested for COVID-19 virus because there is limited equipment to test everyone, while 37.5% suggested that there is limited workforce to test everyone, as shown in Fig. 4.7.

As affirmed by 73.3% of the healthcare worker respondents, it takes more than 1 day to receive COVID-19 test results while 6.7% stated that it only takes 1–2 h to get the results. Still in the same matter, 6.7% suggested that it

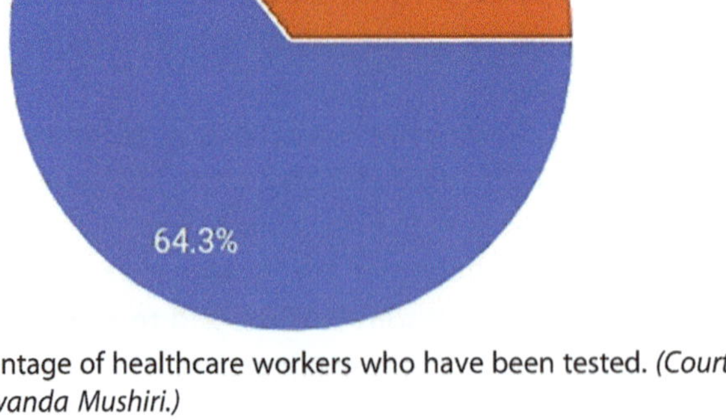

Fig. 4.5 Percentage of healthcare workers who have been tested. *(Courtesy Marvellous Moyo and Tawanda Mushiri.)*

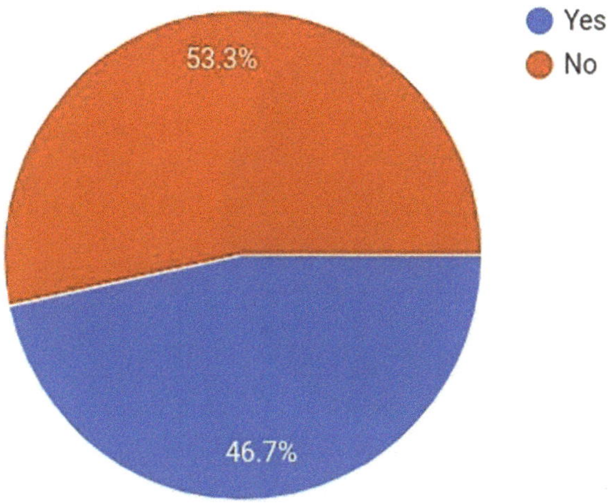

Fig. 4.6 Percentage of individuals that can be tested. *(Courtesy Marvellous Moyo and Tawanda Mushiri.)*

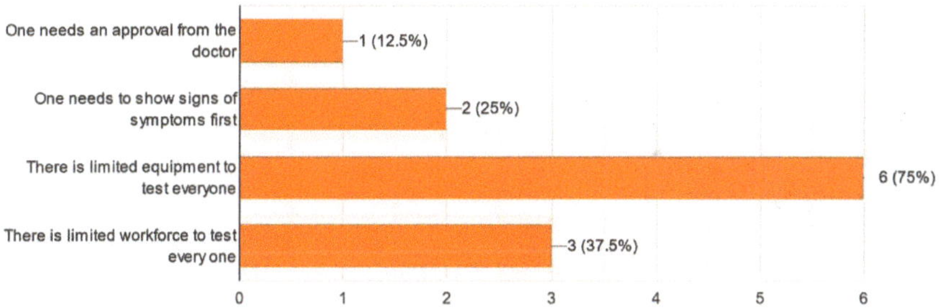

Fig. 4.7 Reasons why not everyone can be tested. *(Courtesy Marvellous Moyo and Tawanda Mushiri.)*

can take 30–60 min to receive the results while the same number of respondents thinks it takes approximately 15 min to receive the results (see Fig. 4.8).

According to 93.3% of the healthcare worker respondents, they are using RT-PCR diagnosis test as a technique for testing COVID-19 virus, while 33.3% stated that they are using antigen diagnosis test and 40% said they are using the antibody testing technique to test for the virus (see Fig. 4.9).

According to the respondents, RT-PCR diagnosis test has the highest accuracy, followed by the antigen test, then the antibody test, as shown in Fig. 4.10. One means very inaccurate and 5 means very accurate.

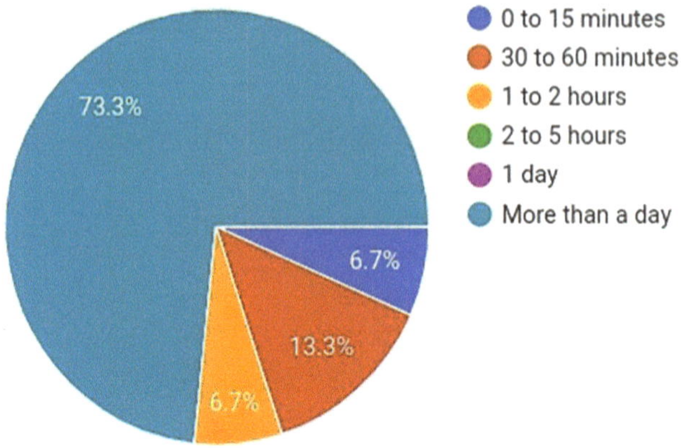

Fig. 4.8 Time taken to receive COVID-19 test results. *(Courtesy Marvellous Moyo and Tawanda Mushiri.)*

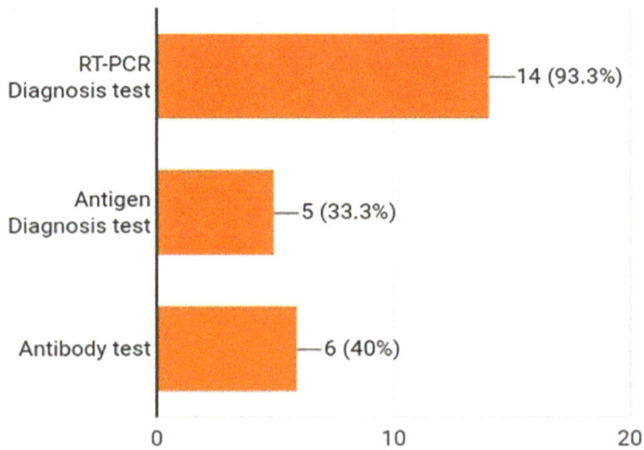

Fig. 4.9 Type of COVID-19 testing techniques. *(Courtesy Marvellous Moyo and Tawanda Mushiri.)*

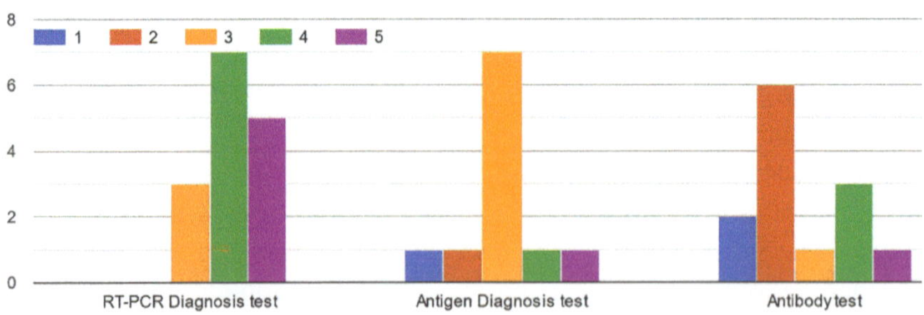

Fig. 4.10 Accuracy of COVID-19 testing techniques. *(Courtesy Marvellous Moyo and Tawanda Mushiri.)*

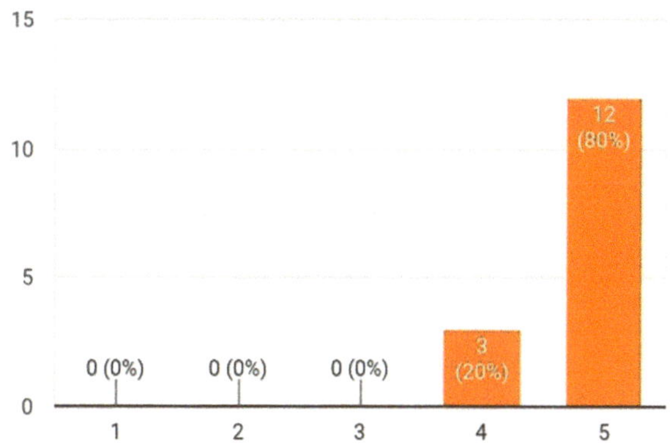

Fig. 4.11 Degree of COVID-19 virus effects on hospitals. *(Courtesy Marvellous Moyo and Tawanda Mushiri.)*

4.5.2 Challenges faced by hospitals during the COVID-19 pandemic

Eighty percent of the healthcare worker respondents stated that the hospital has been extremely affected by the COVID-19 pandemic while 20% said that the effect is high but to the extent affirmed by the former (see Fig. 4.11).

According to all the respondents, there is inadequate supply of personal protective equipment in the hospital while 80% said there were shortages of staff in the hospital. 66.7% of the healthcare workers said the problem was also the shortages of beds and 46.75% suggested one of the challenges the hospital was facing was the inadequate supply of sanitizers. In addition, 6.7% suggested that some other challenges included inadequate investigation techniques and inadequate provision of routine services due to the lack of PPE and fear of the pandemic (see Fig. 4.12).

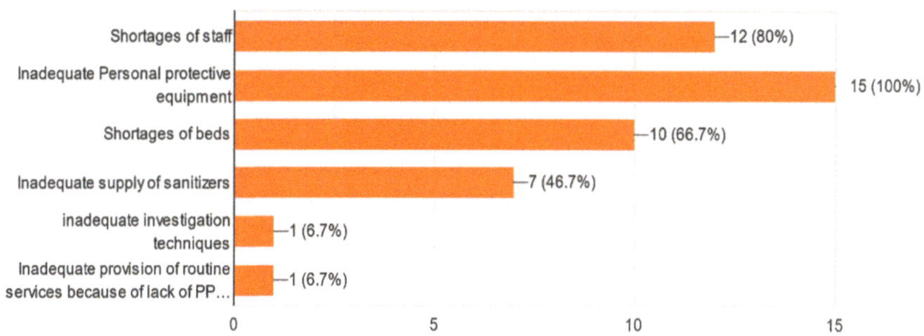

Fig. 4.12 Challenges faced by hospitals. *(Courtesy Marvellous Moyo and Tawanda Mushiri.)*

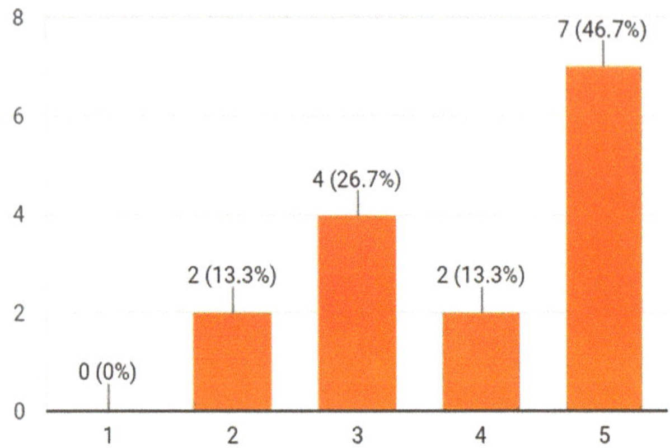

Fig. 4.13 Probability of being infected with COVID-19 virus. *(Courtesy Marvellous Moyo and Tawanda Mushiri.)*

46.7% of healthcare worker respondents stated that they were highly likely to be infected with the virus due to inadequate personal protective equipment. 26.7% said their probability is even lesser while another 13.3% said the chances of being infected with the COVID-19 virus were there but not that high (see Fig. 4.13). One means extremely unlikely while 5 means extremely likely.

4.5.3 Medical use of robots

93.3% of the healthcare workers said they were aware of the use of robots to assist medical doctors in human surgical procedures while 6.7% affirmed that they did not know that robots were used to help doctors in surgical procedures (see Fig. 4.14)

Again, 93.3% of the respondents said they knew what telemedicine is while 6.7% said they did not, as shown in Fig. 4.15.

4.6 Discussion

As an exploration of the current COVID-19 virus testing techniques used in hospitals, this study results show that there are various perspectives regarding the challenges the hospitals are facing during the pandemic and the use of technology in medicine. The findings show that due to lack of man power and resource not everyone can be tested for the COVID-19 virus. This is further strengthened by the fact that some of the healthcare professionals have not yet been tested besides being staff members in this hospital.

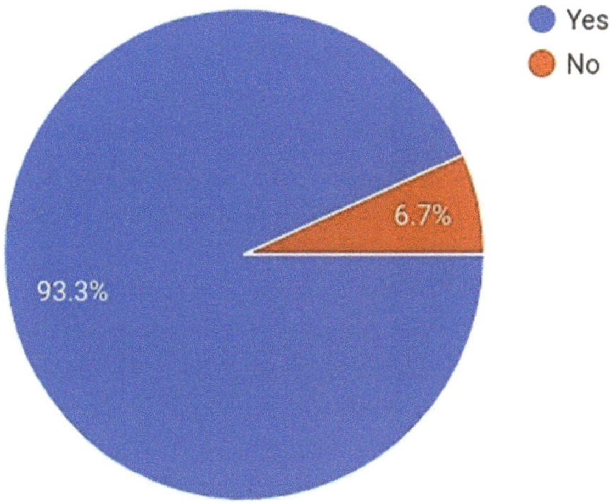

Fig. 4.14 Medical applications of robots. *(Courtesy Marvellous Moyo and Tawanda Mushiri.)*

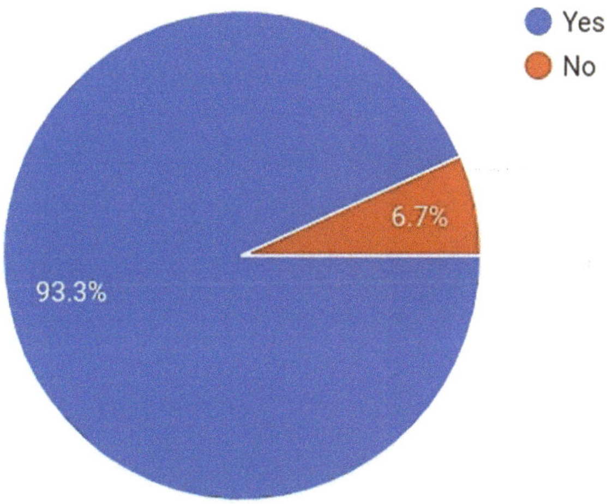

Fig. 4.15 Healthcare workers' knowledge of telemedicine. *(Courtesy Marvellous Moyo and Tawanda Mushiri.)*

Many procedures are needed before one can be tested such as getting approval from the doctor first and to top on that, study findings show that in takes more than 1 day to receive COVID-19 results. From the findings most hospitals are using RT-PCR testing technique, which is very accurate compared to antigen and antibody test in diagnosing COVID-19, but takes a long time to produce results due to a lot of tedious tasks involved. This confirms the literature survey findings, which affirms that RT-PCR takes approximately 5 h to produce the test results (Gignoux et al., 2016; Pascarella et al., 2020). In addition, the shortages of staff in hospitals also aid to the delay of the testing results and indirectly resulting in escalation of COVID-19 cases. To lessen the spread of COVID-19 virus it is crucial that an enormous number of individuals should be tested and hence there is a need to find systems that can help address that imminently.

Results show that COVID-19 pandemic affected the hospitals extremely badly and the challenges that hospitals are facing are mainly the shortages of staff, resources, and appropriate technology to battle the pandemic. The current COVID-19 testing and sanitization techniques used in hospitals not only do they require continuous changing of PPE but they put the lives of healthcare workers at higher risk of infection. This is shown by the result findings, which affirmed that the probability of healthcare professionals being affected is very high. Similarly, the literature findings show that the medical care workers are at a higher risk of infection while at the same time the hospitals are facing resource shortages (Gao & Xu, 2008; Wang et al., 2020). Thus there is a need to look into current technology, how it can be adapted to improve the testing process of COVID-19 virus and help reduce the risks of infection of healthcare workers. Though a lot of healthcare professionals are aware of the use of robots in medicine and telemedicine, still such technology has not yet been fully exhausted in regard to testing COVID-19 virus. It is against this background that the researchers proposed the design of an intelligent robotic testing machine for testing COVID-19 virus to be installed in hospitals.

4.7 Conclusion

According to the World Health Organization (WHO), the novel coronavirus COVID-19 is a global pandemic which is a very dangerous threat to human and testing is very essential in managing the propagation of this virus. This is because testing helps determines who needs isolation from other people and whom to distance oneself from. A survey was done to determine the perspectives of healthcare workers with regard to COVID-19 testing and ask them about the challenges the hospital was facing during the pandemic. It was shown from the research findings that the hospital was facing a lot of challenges, with shortage of equipment and staff being on top of the list. Owing to these challenges and ineffective testing techniques, the time taken to receive COVID-19 test results is too long and hence a better approach is needed to address these concerns. The researchers have suggested the design of an intelligent robotic machine for testing COVID-19 virus which will be installed in all the hospitals. Also, this study is part of an ongoing research on the design of the just mentioned machine and will contribute mainly on the justification of why there is a need to incorporate such a machine in the hospitals.

References

Alexandrovsky, D., et al. (2020). Examining design choices of questionnaires in VR user studies. In *Conference on human factors in computing systems—Proceedings*Association for Computing Machinery. https://doi.org/10.1145/3313831.3376260.

Avert. (2020). *Infographics on coronavirus (COVID-19)*. Avert. Available at: https://www.avert.org/coronavirus/infographics (Accessed 14 August 2020).

Beeching, N., Fletcher, T., & Beadsworth, M. (2020). *Rapid near patient testing for both current and past infections is urgently required*. https://doi.org/10.1016/S0140-6736(20)30183-5.

Berg, S. (2009). *Identity theft causes, correlates, and factors: A content analysis*. studydaddy.com. Available at: https://studydaddy.com/attachment/47521/ka18k1f536.pdf (Accessed 23 June 2021).

Bhatraju, P., & Ghassemieh, B. (2020). *Covid-19 in critically ill patients in the Seattle region—case series*. Massachusetts Medical Society. Available at: https://www.nejm.org/doi/full/10.1056/NEJMoa2004500 (Accessed 24 July 2020).

Chakraborty, T., Jamal, R., & Battineni, G. (2020). *A review of prolonged post-COVID-19 symptoms and their implications on dental management*. mdpi.com. Available at: https://www.mdpi.com/1106536 (Accessed 23 June 2021).

Creswell, J. W. (2011). *The SAGE handbook of qualitative research*. Google Books. Available at: https://books.google.co.zw/books?hl=en&lr=&id=qEiC-_ELYgIC&oi=fnd&pg=PA269&dq=Creswell,+2011+research+design&ots=C4jUxlIw5D&sig=CwXxlOlyeD43u40JXUB7KoiQlqo&redir_esc=y#v=onepage&q=Creswell%2C2011researchdesign&f=false (Accessed 23 June 2021).

Ellis, G. (2020). *Pros and cons of COVID-19 testing*. News | phillytrib.com. Available at: https://www.phillytrib.com/news/pros-and-cons-of-covid-19-testing/article_adcffaf6-2cfb-5979-83f5-fe99317313aa.html (Accessed 14 August 2020).

FDA. (2020). *Coronavirus testing basics*. FDA. Available at: https://www.fda.gov/consumers/consumer-updates/coronavirus-testing-basics (Accessed 14 August 2020).

Gandhi, M., Yokoe, D., & Havlir, D. (2020a). *Asymptomatic transmission, the Achilles' heel of current strategies to control COVID-19*. Available at: https://www.nejm.org/doi/full/10.1056/NEJMe2009758 (Accessed 24 July 2020).

Gandhi, M., Yokoe, D. S., & Havlir, D. V. (2020b). Asymptomatic transmission, the achilles' heel of current strategies to control Covid-19. *New England Journal of Medicine*, 2158–2160. https://doi.org/10.1056/NEJMe2009758. Massachussetts Medical Society.

Gao, L. Z., & Xu, H. (2008). Comparisons of mutation rate variation at genome-wide microsatellites: Evolutionary insights from two cultivated rice and their wild relatives. *BMC Evolutionary Biology*, *8*(1). https://doi.org/10.1186/1471-2148-8-11.

Garriga, M., et al. (2020). The role of mental health home hospitalization care during the COVID-19 pandemic. *Acta Psychiatrica Scandinavica*, 479–480. https://doi.org/10.1111/acps.13173. Blackwell Publishing Ltd.

Gignoux, E., et al. (2016). Effect of artesunate–amodiaquine on mortality related to Ebola virus disease. *New England Journal of Medicine*, *374*(1), 23–32. https://doi.org/10.1056/nejmoa1504605.

Healthline. (2020). *2019 coronavirus (COVID-19): Symptoms, treatment, and prevention*. Available at: https://www.healthline.com/health/coronavirus-covid-19#outlook (Accessed 24 July 2020).

Hollander, J. E., & Carr, B. G. (2020). Virtually perfect? Telemedicine for Covid-19. *New England Journal of Medicine*, *382*(18), 1679–1681. https://doi.org/10.1056/NEJMp2003539. Massachusetts Medical Society.

Israfil, S. M. H., et al. (2021). Clinical characteristics and diagnostic challenges of COVID−19: An update from the global perspective. *Frontiers in Public Health*. https://doi.org/10.3389/fpubh.2020.567395. Frontiers Media S.A.

Jang, S., Han, S. H., & Rhee, J. Y. (2020). Cluster of coronavirus disease associated with fitness dance classes, South Korea. *Emerging Infectious Diseases*, *26*(8), 1917–1920. https://doi.org/10.3201/eid2608.200633. Centers for Disease Control and Prevention (CDC).

Kent, C. (2020). *What are the different types of Covid-19 test and how do they work?*. Available at: https://www.medicaldevice-network.com/features/types-of-covid-19-test-antibody-pcr-antigen/ (Accessed 14 August 2020).

Khyber, S. F., et al. (2020). Impacts of climate change on livelihood of older people in district Nowshera-Pakistan view project critical review of past literature of militancy impact on educational institutions in the world view project. *SSRN Electronic Journal*. https://doi.org/10.2139/ssrn.3655613.

Kim, P. S. (2020). South Korea's fast response to coronavirus disease: Implications on public policy and public management theory. *Public Management Review*. https://doi.org/10.1080/14719037.2020.1766266. Taylor and Francis Ltd.

Lan, L., et al. (2020a). *Positive RT-PCR test results in patients recovered from COVID-19*. jamanetwork.com. Available at: https://jamanetwork.com/journals/jama/article-abstract/2762452 (Accessed 24 July 2020).

Lan, L., et al. (2020b). Positive RT-PCR test results in patients recovered from COVID-19. *JAMA: The Journal of the American Medical Association*, 1502–1503. https://doi.org/10.1001/jama.2020.2783. American Medical Association.

Lincoln, Y., & Guba, E. G. (1985). *Naturalistic inquiry*. Newbury Park, CA: Sage.

Morse, J. M., et al. (2002). Verification strategies for establishing reliability and validity in qualitative research. *International Journal of Qualitative Methods*, *1*(2), 13–22. https://doi.org/10.1177/160940690200100202. SAGE Publications.

Pascarella, G., et al. (2020). COVID-19 diagnosis and management: A comprehensive review. *Journal of Internal Medicine*, 192–206. https://doi.org/10.1111/joim.13091. Blackwell Publishing Ltd.

Patient Power. (2020). *The pros and cons of the COVID-19 antibody tests*. Patient Power. Available at: https://patientpower.info/navigating-cancer/support-and-resources/the-pros-and-cons-of-the-covid-19-antibody-tests (Accessed 14 August 2020).

Prabhakar, P. K., & Lakhanpal, J. (2020). Recent advances in the nucleic acid-based diagnostic tool for coronavirus. *Molecular Biology Reports*, 9033–9041. https://doi.org/10.1007/s11033-020-05889-3. Springer Science and Business Media B.V.

Pyett, P. M. (2003). Validation of qualitative research in the "real world". *Qualitative Health Research*, *13*(8), 1170–1179. https://doi.org/10.1177/1049732303255686.

Radley, A., & Chamberlain, K. (2012). The study of the case: Conceptualising case study research. *Journal of Community and Applied Social Psychology*, *22*(5), 390–399. https://doi.org/10.1002/casp.1106.

Ray, D., & Subramanian, S. (2020). India's lockdown: An interim report. *Indian Economic Review, 55*, 31–79. https://doi.org/10.1007/s41775-020-00094-2. Editorial Office, Indian Economic Review.

Reynolds, H., & Adhikari, S. (2020). *Renin–angiotensin–aldosterone system inhibitors and risk of Covid-19*. Massachusetts Medical Society. Available at: https://www.nejm.org/doi/full/10.1056/NEJMoa2008975 (Accessed 24 July 2020).

Sen, A. P., & Rout, N. K. (2021). Implementation of transfer learning technique for the detection of COVID-19. In *Lecture notes in electrical engineering* (pp. 135–140). Springer Science and Business Media Deutschland GmbH. https://doi.org/10.1007/978-981-33-4866-0_17.

Shaker, M. S., et al. (2020). COVID-19: Pandemic contingency planning for the allergy and immunology clinic. *The Journal of Allergy and Clinical Immunology: In Practice*, 1477–1488. e5. https://doi.org/10.1016/j.jaip.2020.03.012. American Academy of Allergy, Asthma and Immunology.

Singh, K., Fox, J. R. E., & Brown, R. J. (2016). Health anxiety and Internet use: A thematic analysis. *Cyberpsychology: Journal of Psychosocial Research on Cyberspace, 10*(2), 4. https://doi.org/10.5817/CP2016-2-4.

Sohrabi, C., et al. (2020). World Health Organization declares global emergency: A review of the 2019 novel coronavirus (COVID-19). *International Journal of Surgery*, 71–76. https://doi.org/10.1016/j.ijsu.2020.02.034. Elsevier Ltd.

Tasdemir, A. (2010). *The content analysis of the news in the national papers concerning the renewed primary curriculum*. ERIC. Available at: https://eric.ed.gov/?id=EJ919895 (Accessed 23 June 2021).

van Doremalen, N., et al. (2020). Aerosol and surface stability of HCoV-19 (SARS-CoV-2) compared to SARS-CoV-1. *medRxiv: The preprint server for health sciences*. https://doi.org/10.1101/2020.03.09.20033217.

Vieta, E., Pérez, V., & Arango, C. (2020). Psychiatry in the aftermath of COVID-19. *Revista de Psiquiatría y Salud Mental (English Edition), 13*(2), 105–110. https://doi.org/10.1016/j.rpsmen.2020.04.004. Elsevier BV.

Wang, C., Ng, C., & Jama, R. B. (2020). *Response to COVID-19 in Taiwan: Big data analytics, new technology, and proactive testing*. jamanetwork.com. Available at: https://jamanetwork.com/journals/jama/article-abstract/2762689 (Accessed 24 July 2020).

Wang, X., et al. (2020). *Association between universal masking in a health care system and SARS-CoV-2 positivity among health care workers*. jamanetwork.com. Available at: https://jamanetwork.com/journals/jama/article-abstract/2768533 (Accessed 23 June 2021).

Watkins-Hayes, C. (2019). 1. Dying from: Sexual violence, the drug economy, and the persistence of HIV/AIDS. In *Remaking a life* (pp. 35–79). University of California Press. https://doi.org/10.1525/9780520968738-004.

WHO. (2021). *WHO Coronavirus (COVID-19) Dashboard | WHO Coronavirus (COVID-19) dashboard with vaccination data*. Available at: https://covid19.who.int/ (Accessed 22 June 2021).

Wu, E. (2020). *Antigen tests for COVID-19 are fast and easy – and could solve the coronavirus testing problem despite being somewhat inaccurate*. Available at: https://theconversation.com/antigen-tests-for-covid-19-are-fast-and-easy-and-could-solve-the-coronavirus-testing-problem-despite-being-somewhat-inaccurate-137977 (Accessed 14 August 2020).

Yin, R. K. (2013). Validity and generalization in future case study evaluations. *Evaluation, 19*(3), 321–332. https://doi.org/10.1177/1356389013497081.

Yin, T. M., et al. (2009). Microsatellite primer resource for Populus developed from the mapped sequence scaffolds of the Nisqually-1 genome. *New Phytologist, 181*(2), 498–503. https://doi.org/10.1111/j.1469-8137.2008.02663.x. John Wiley & Sons, Ltd.

Zoghbi, W. A., et al. (2020). Multimodality cardiovascular imaging in the midst of the COVID-19 pandemic: Ramping up safely to a new normal. *JACC: Cardiovascular Imaging, 13*(7), 1615–1626. https://doi.org/10.1016/j.jcmg.2020.06.001. Elsevier Inc.

Further reading

Danilyuk, P. (2021). *Black and silver steel tool · free stock photo.* Available at: https://www.pexels.com/photo/black-and-silver-steel-tool-8438864/ (Accessed 14 September 2021).

CHAPTER 5

Development of an intelligent specimen testing booth to reduce healthcare professionals' infection risks

5.1 Introduction

COVID-19 virus has exposed healthcare workers and their families to unprecedented degrees of risk (Temsah et al., 2020). Albeit not representative, information from numerous nations across WHO regions shows that COVID-19 virus infections among healthcare workers are far more noteworthy than those in the general population.

While healthcare workers represent under 3% of the population in the vast larger part of nations and under 2% in practically all low- and middle-income nations, around 14% of COVID-19 virus cases reported to WHO are among healthcare workers. In certain nations, the extent can pretty much be as high as 35% (Pappa et al., 2020). Nonetheless, information accessibility and quality are restricted, and it is impossible to establish whether healthcare workers were infected in the work place or in local area settings. A huge number of medical care workers infected with COVID-19 virus have lost their lives around the world.

Notwithstanding physical risks, the pandemic has put uncommon degrees of mental stress on healthcare workers exposed to high-demand settings for extended periods, living in consistent dread of infection exposure while isolated from family, and confronting social slander (Taylor Professor of Work, 2020). Before COVID-19 virus hit, medical experts were already at higher danger of suicide in all places of the world (Mortier et al., 2020; Pearman et al., 2020; Ranney, Griffeth, & Jha, 2020). WHO as of late featured a disturbing ascent in reports of verbal discrimination, harassment, and actual brutality among medical care workers in the wake of COVID-19 virus (Borrion et al., 2020; Rohwerder, 2020).

Healthcare Systems Design of Intelligent Testing Centers
https://doi.org/10.1016/B978-0-323-99443-9.00002-4

Keeping medical experts sound during the COVID-19 pandemic is fundamental in both easing back the pace of infection and meeting uplifted staffing needs (Gössling, Scott, & Hall, 2020; Vest, Blackburn, & Yeager, 2020). Numerous legislatures and medical care suppliers are finding this to be a troublesome assignment because of the worldwide lack of personal protective equipment (PPE) supplies.

COVID-19 virus testing activities are frequently the first physical point of contact between a medical care provider and patient, carrying with it an innate danger of transmission (Driggin et al., 2020; Pfaar et al., 2021; Su et al., 2020). Enormous drive-through testing tasks have demonstrated compelling in restricting provider-to-patient exposure and speeding up test administration. Yet, this type of testing requires critical PPE supplies for testing center staff and administrators. It likewise presents accessibility challenges for communities where enormous fragments of the populace do not have access to a vehicle. A walk-in testing booth gives an elective arrangement that eliminates healthcare worker to patient exposure in a modular configuration that is easy to deploy for brief testing tasks.

5.2 Literature review

The University of Maryland researchers designed a booth, made essentially of Perspex and aluminum, that uses a HEPA (high efficiency particulate air) filter blower to make a positive pressure system (University of Maryland, 2020). This shields the people performing the test by forestalling virus particles and different germs from entering the booth. Along these lines, the booth establishes a cleanroom environment similar to those generally utilized in the pharmaceutical and healthcare device industries to test items and medical apparatuses without the danger of contaminants. For the testing technique, a patient sits outside the booth in a seat situated close to the gloves. The healthcare worker inside the booth utilizes the gloves to recover a testing swab from a plate mounted outwardly. Then, the healthcare worker utilizes a radio framework to request that the patient slant their head, prior to utilizing the swab to gather a specimen from the cavity between the patient's nose and mouth. The healthcare worker then collects the swab through a plate attached outwardly of the booth. When the patient leaves the region, test site personnel completely clean the gloves, seat, and front side of the booth. Inside the booth, when they are now prepared to receive the following patient, the person steps on a pneumatic switch to show a sign.

The Hogan Lab research group, alongside ME staff individuals, constructed game-changing booth with positive pressure HEPA filtered air to permit medical service workers to securely test COVID-19 patients without the requirement of PPE (University of Minnesota, 2020). The booth permits medical service workers to tackle the issue of PPE deficiency by establishing an environment where PPE is not needed. Regularly, medical service providers would have to change their mask, outfit, and gloves between each test. In a booth, this is no longer a necessity, so providers can test twice as numerous individuals each day. Rather than exchanging PPE between tests, the only disinfecting that needs to happen is on the exterior of the booth. The booth likewise has a fan and HEPA channel that establishes a positive pressure environment that repulses outside particulates, keeping the medical service provider in the booth safe. The booth permits healthcare services to increase testing while at the same time saving PPE for individuals in hospital who need it most.

The rise of COVID-19 in South Korea, and the private and public sector reaction to it, fills in as an important contextual analysis for nations confronting comparable outbreaks (Lee & Lee, 2020). This chapter centers around how Korean health authorities carried out drive and walk diagnostic testing, and broad movement and contact tracing, to distinguish and inform exposed individuals of the general population. This contextual analysis shows the significance of quick variation of transportation infrastructure and location-based information technology to react to public health emergencies, and how governments can learn from experimentation and past experience to speed up these responses.

According to Labline (2020) by making a completely encased, positive pressure environment inside the booth and the execution of HEPA filtration, tainting hazard for the medical care specialist is decreased. The Garmat Walk-Up testing booth tends to a few weaknesses of current testing strategies. In the general medical setting, when a patient is being tested for the COVID-19 infection, the medical provider should wear PPE. The PPE should be changed after every patient, paying little attention to result of the test. While a couple of drive-through testing stations have been set up, they are tedious and not practical for those without access to transportation. In related news, Brigham and Women's Hospital in Boston developed a three-sided plexiglass version, an advancement over a South Korea booth, as it put the medical care provider inside, limiting the surface region and time needed for sterilization. Garmat's design improves Brigham's version to give extra security to medical care workers by completely encasing the booth and applying positive air pressure flow, forestalling contaminants including the infection, from entering the inside of the booth.

5.3 Methods and materials

The intelligent robot machine should be able to test patients without any human supervision. Also, the system should be able to test everyone who will be entering the hospital. In order to achieve this, an automatic booth will be placed at the entrance. The booth will mainly consist of the robotic manipulator, the testing kits such as the swabs, chemicals, and lastly a table where the robotic manipulator will be attached.

5.3.1 Determining the anthropometry

To design the booth effectively, the standard size of the human body is of utmost importance. In this report, the main focus is to determine an adult human being's standard height and weight. For this book, the standard height of the human being has been assumed to be 2 m, that is, for simplicity's sake and the weight is assumed to be 70 kg. Note should be taken that a study can be done to determine the average values for a given sample of people but this is beyond the project scope.

5.3.2 Intelligent COVID-19 testing center

Fig. 5.1 shows each component of the proposed booth design. A dual-compartment enclosure is for the design. Therefore there is a physical barrier between the intelligent robotic machines during swab collection and patient being tested. The first compartment with a door is for the patient and has 5.42 m^2 area and 2 m height, maintained at negative pressure to reduce spreading of the virus to the external environment. This compartment consists of a chair, LCDs, and speakers. The speakers and LCDs will be used to guide the patient to sit on the chair and at the same time educate them about how the testing will be conducted. All the equipment necessary for swab collection and testing are placed in the compartment with the robotic testing machines. The partition between the robotic machines and patient can slide up and down during the testing and after the testing processes, respectively. Also, the robotic machine's compartment is equipped with tables to improve the working conditions. The booth frame is fabricated using aluminum extrusions (40 mm × 40 mm nominal cross-section). The walls are made of Perspex with 5 mm nominal thickness. Two 230 V, 8 W LED lamps

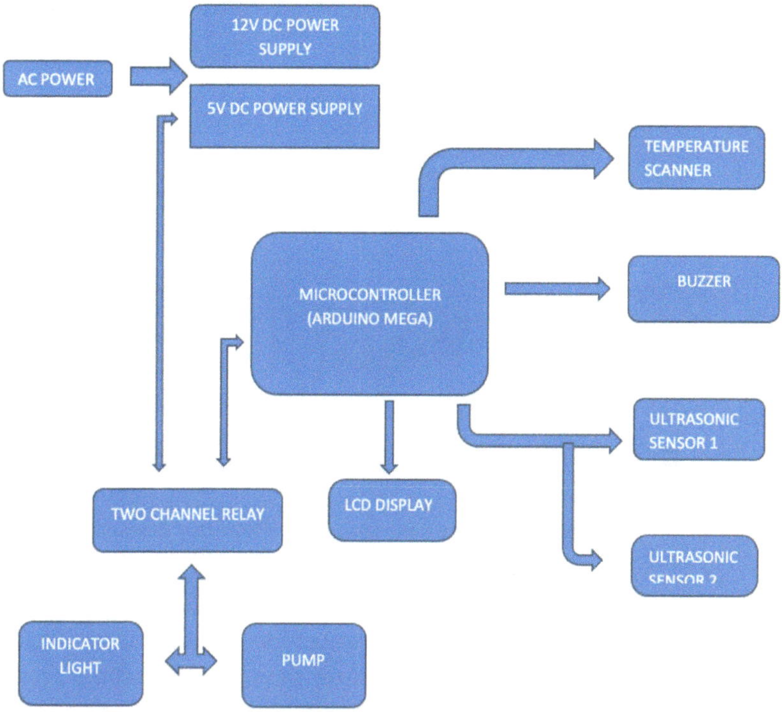

Fig. 5.1 Components of the intelligent booth. *(Courtesy Marvellous Moyo and Tawanda Mushiri.)*

are used for interior lighting. Patient's compartment top of 5.42 m^2 area and 0.565 m height consists of a High-Efficiency Particulate Air filter (HEPA) (0.3 μm nominal cut of) with 100–150 m^3/h airflow capacity against 50 Pa differential pressure, a set of four UV-C lamps (254 nm, 30 W each), and a conventional blower (100–150 m^3/h) matched to the HEPA filter to generate the pressure differential in the chamber.

This system is used to disinfect the exhaust air and trapping the fine viral matter. Volume of air in the top compartment gets exposed for at least 20 s before it reaches to HEPA filter. This is done to make sure that the major part of suspended viral load is sterilized before it reaches the filter and the HEPA filter aids this disinfection. Prior to this, automatic electrostatic sprayers with chemical disinfectant are used to disinfect the testing robotic arm, the chair, and the patient compartment. Automatic disinfection is repeated after every patient exits. The robotic machine's compartment is equipped with four UV lamps (254 nm, 30 W each) for continuous

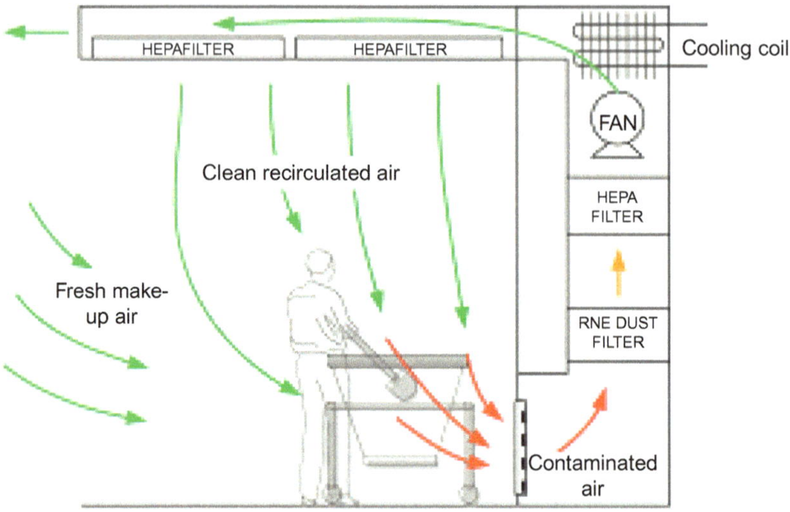

Fig. 5.2 Airflow inside the booth. *(Courtesy Marvellous Moyo and Tawanda Mushiri.)*

disinfection but not during testing. An air blower is used to maintain a positive pressure in the robotic machine's chamber. So, during testing when the partition is opened, air can only move from the robotic machine's testing side to the patient's side, thereby further preventing contamination of the testing equipment. The same principle of the HEPA system is used but now air will be moving from outside to the robotic machine's chamber then through partition to the patient's chamber, after which is exhausted to the external environment (see Fig. 5.2). Since the whole testing process is fully autonomous, a camera is installed outside the door which will be used for identifying the patient's face and then the door is opened. Both the partition and door consist of a motor and a rack and pinion system. Through this, both the door and partition can be opened and closed. The whole system is controlled using a raspberry pie.

5.3.3 Negative pressure inside the booth

A M2 Series Smart Sensor was utilized for pressure variation measurement. A small hole was drilled on the Perspex wall and one limb of the manometer was held in the booth interior while the other was routed externally. Two manometers can be used, one in the robotic machine's testing chamber and the other on patient's chamber. Measurements can be taken with the booth door closed and blower fan switched on. The booth is made air tight, so there is no leakage of air to the outside environment.

5.4 Results and discussion

The booth will be designed based on the information obtained in the anthropometry study. From that study the average height of a human being is 2 m and the two robotic arms are both approximately 1.2 m long with a spherical workspace. Therefore, from this data, the booth will be rectangular prism with dimensions given as 3 m × 3 m × 2.5 m. Perspex has been selected as the main material for the door and side walls while aluminum will be used as the frame material. The earlier figure shows the requisite components of the booth. All components of the Intelligent Robotic Testing machine will be enclosed inside the booth.

According to the International Ultraviolet Association, the UV dose of 40 mJ/cm^2 should be used for disinfection of viruses on flat, ideal surfaces (Derraik et al., 2020; International Ultraviolet Association, 2021). With the configuration of this booth design this power density can be accumulated in 3 min. The exhaust system utilized in the booth must maintain a negative pressure differential of at least 12 Pa, which satisfies the guidelines set by the Centers for Disease Control and Prevention in 2003 (Sehulster & Chinn, 2003).

The design of an intelligent booth aims to protect healthcare workers from the risk of infection during swab collection for testing COVID-19 virus. Negative pressure is one of the essential safety measures which can be employed in the design of booth. This can be achieved through the use of HEPA and blower fan. To aid safety, UV lamps and automatic electrostatic sprayers can be used. To control the opening and closing of booth, rack and pinion gear system driven by a motor can be used. The whole system can be controlled using a raspberry pie, which also controls the robotic manipulator used for sample collection and testing. The utilization of such systems can minimize the need of a continuous change of PPEs during COVID-19 sample collection, thus minimizing the use of PPEs as well as risk that healthcare workers have to undergo.

5.5 Recommendations and conclusion

It is recommended that future researches need to be done on effective methods for disinfecting contaminated surfaces. Also, since there is more room in the patient's chamber and two robotic arms capable of testing, a study can be done on how the booth can be redesigned to test more than one patient. During COVID-19 virus testing, the healthcare workers are at a higher risk of being infected with COVID-19 virus. It is against this background that the researchers have decided to design an intelligent booth used for testing COVID-19 virus without human intervention. In this study, development consideration necessary for an intelligent booth for testing COVID-19 virus has been done. It has been found out that the use of systems like robotic manipulators, cameras for facial identification, rack and pinion motor driven gears, HEPA, UV lamps, and electrostatic sprayers with disinfectant can be used to test patients without the need of healthcare professionals. The study findings contribute to the ongoing design of an intelligent machine for testing COVID-19 virus, by providing information that will be used for controlling the robotic manipulators during sample collection.

References

Borrion, H., et al. (2020). Measuring the resilience of criminogenic ecosystems to global disruption: A case-study of COVID-19 in China. *PLoS One*, *15*(10 October). https://doi.org/10.1371/journal.pone.0240077. Public Library of Science.

Derraik, J. G. B., et al. (2020). Rapid review of SARS-CoV-1 and SARS-CoV-2 viability, susceptibility to treatment, and the disinfection and reuse of PPE, particularly filtering facepiece respirators. *International Journal of Environmental Research and Public Health*, *17*(17), 1–31. https://doi.org/10.3390/ijerph17176117. MDPI AG.

Driggin, E., et al. (2020). Cardiovascular considerations for patients, health care workers, and health systems during the COVID-19 pandemic. *Journal of the American College of Cardiology*, 2352–2371. https://doi.org/10.1016/j.jacc.2020.03.031. Elsevier USA.

Gössling, S., Scott, D., & Hall, C. M. (2020). Pandemics, tourism and global change: A rapid assessment of COVID-19. *Journal of Sustainable Tourism*, 1–20. https://doi.org/10.1080/09669582.2020.1758708. Taylor & Francis. Routledge https://www.tandfonline.com/loi/rsus20.

International Ultraviolet Association. (2021). *International Ultraviolet Association Inc—Home*. Available at: https://iuva.org/ (Accessed 1 June 2021).

Labline. (2020). *Walk-up testing booth protects medical workers against COVID–19 virus*. Google Search. Available at: https://www.mlo-online.com/disease/infectious-disease/article/21140914/walkup-testing-booth-protects-medical-workers-against-covid19-virus.

Lee, D., & Lee, J. (2020). *Testing on the move: South Korea's rapid response to the COVID-19 pandemic*. Elsevier. Available at: https://www.sciencedirect.com/science/article/pii/S2590198220300221 (Accessed 1 June 2021).

Mortier, P., et al. (2020). Thirty-day suicidal thoughts and behaviors among hospital workers during the first wave of the Spain COVID-19 outbreak. *Depression and Anxiety*, *22*, 36. https://doi.org/10.1002/da.23129 (Blackwell Publishing Inc.) https://pubmed.ncbi.nlm.nih.gov/33393724/10.1002/da.23129.

Pappa, S., et al. (2020). Prevalence of depression, anxiety, and insomnia among healthcare workers during the COVID-19 pandemic: A systematic review and meta-analysis. *Brain, Behavior, and Immunity*, *88*, 901–907. https://doi.org/10.1016/j.bbi.2020.05.026. Academic Press Inc.

Pearman, A., et al. (2020). Mental health challenges of United States healthcare professionals during COVID-19. *Frontiers in Psychology, 11*. https://doi.org/10.3389/fpsyg.2020.02065. Frontiers Media S.A.

Pfaar, O., et al. (2021). COVID-19 pandemic: Practical considerations on the organization of an allergy clinic—An EAACI/ARIA Position Paper. *Allergy: European Journal of Allergy and Clinical Immunology, 76*(3), 648–676. https://doi.org/10.1111/all.14453. Blackwell Publishing Ltd.

Ranney, M. L., Griffeth, V., & Jha, A. K. (2020). Critical supply shortages—The need for ventilators and personal protective equipment during the Covid-19 pandemic. *New England Journal of Medicine, 382*(18), e41. https://doi.org/10.1056/nejmp2006141. Massachusetts Medical Society.

Rohwerder, B. (2020). *Emerging issues report social impacts and responses related to COVID-19 in low- and middle-income countries*. Available at: https://opendocs.ids.ac.uk/opendocs/handle/20.500.12413/15625 (Accessed 1 June 2021).

Sehulster, L., & Chinn, R. Y. W. (2003). *Guidelines for environmental infection control in health-care facilities: Recommendations of CDC and the Healthcare Infection Control Practices Advisory Committee (HICPAC)*. Available at: https://www.cdc.gov/mmwr/preview/mmwrhtml/rr5210a1.htm (Accessed 1 June 2021).

Su, L., et al. (2020). The different clinical characteristics of corona virus disease cases between children and their families in China-the character of children with COVID-19. *Emerging Microbes & Infections, 9*(1), 707–713. https://doi.org/10.1080/22221751.2020.1744483. Taylor and Francis Ltd.

Taylor Professor of Work. (2020). *Covid-19: Contact/call centre workers in Scotland*. Available at: https://strathprints.strath.ac.uk/id/eprint/72600 (Accessed 1 June 2021).

Temsah, M. H., Al-Sohime, F., Alamro, N., Al-Eyadhy, A., Al-Hasan, K., Jamal, A., et al. (2020). The psychological impact of COVID-19 pandemic on health care workers in a MERS-CoV endemic country. *Journal of Infection and Public Health, 13*(6), 877–882. https://doi.org/10.1016/j.jiph.2020.05.021. Elsevier Ltd.

University of Maryland. (2020). *Researchers design mobile COVID-19 testing booths*. A. James Clark School of Engineering, University of Maryland. Available at: https://eng.umd.edu/news/story/researchers-design-mobile-covid19-testing-booths (Accessed 1 June 2021).

University of Minnesota. (2020). *ME helps design COVID-19 testing booth, allowing for twice as many tests per day*. Mechanical Engineering | College of Science and Engineering. Available at: https://cse.umn.edu/me/news/me-helps-design-covid-19-testing-booth-allowing-twice-many-tests-day (Accessed 1 June 2021).

Vest, J. R., Blackburn, J., & Yeager, V. A. (2020). Challenges in translating national and state reopening plans into local reopening policies during the COVID-19 pandemic. National Response and Reopening Plans Cannot Be Adopted Wholesale *Public Health Reports*. https://doi.org/10.1177/0033354920971721. SAGE Publications Ltd.

CHAPTER 6

Using robotics, artificial intelligence, and deep learning to collect COVID-19 samples

6.1 Introduction to artificial intelligence (AI)

The coming of artificial intelligence and its advancement has drawn in publicity of conversation in the previous few years because of its quickly expanding contributions under different domains (Dimiduk, Holm, & Niezgoda, 2018). The idea of AI can be clarified under the light of Ramon Llull's hypothesis of thinking machine and Aristotle's arguments which were created in CE 1300 and 300 BC, respectively (Press, 2020). By the progression of time for example in the 1950s, a reasonable and pragmatic definition with respect to the AI innovation was proposed (Flogie & Aberšek, 2019; Skyttner, 2001). In the current time, the role and proficiency of AI has gone through an emotional restoration because of the broad contribution of innovative work in the field of artificial intelligence programs. Major worldwide enterprises have made tremendous interests into the AI programming and the development of innovative AI applications and programming (He et al., 2019; Makridakis, 2017).

Numerous instruments have been utilized by artificial intelligence to automate the problem-based assignments. Such devices are being created following the AI standards and are utilized to make diverse artificial intelligence-based applications that are significant in settling issues across various disciplines and enterprises (Singh Gill et al., 2019; Singh, Sharma, & Yoon, 2020). Search in this sort of framework mirrors the genuine real-life problem-based arrangements by the use of computing power to determine these issues. These search problems are generally ordered dependent on the measure of data introduced to the search cycle. Such a data is related either to the general problem area or to any particular piece of the program. This sort of search and advancement depends on various ordinary methods.

It is critical to pinpoint the viability of the search cycle which makes it novel and unique in relation to other significant innovations right now

https://doi.org/10.1016/B978-0-323-99443-9.00003-6

working in comparable area. The ordinary strategy adopted by the artificial intelligence incorporates past search findings, which AI learns and utilizes to refine its search cycle in contrast with the earlier searches and along these lines plan forward while explaining the significant questions like the human knowledge (Keding, 2020; Yang et al., 2019). The part of evolutionary computation is one such model and is alluded as the umbrella term for calculations following the public evolutionary cycles which fuses the system of automized selection, while giving endurance to the fittest principle.

6.1.1 AI and medicine

Considering the viability of the artificial intelligence implications in pretty much every department, a huge consideration has been allowed toward its application in the medical care department. Distinctive artificial intelligence-based arrangements are being utilized to improve the adequacy in the conveyance of medical care while tending to helpful solutions and data in regard to the beforehand interactable issues (Nammour, 2018; Rojo, 2020). Following this, few artificial intelligence-based applications and other smart arrangements have been acquainted with the medical care market, which fills in as the tribute to this expanded attention. Various analysts and medical services specialists have examined how the utilization of artificial intelligence innovation in the medical care department will additionally contribute sooner rather than later (Rojo, 2020).

A huge expansion in the execution of artificial intelligence innovation has been seen in different clinical settings, since the cutting-edge medication is confronting different difficulties in extracting, examining, and applying both organized and unstructured information to guarantee illness management through its special information mining as well as acknowledgment capacity (Nammour, 2018; Yang et al., 2019). With this in mind, clinical artificial intelligence is usually related to the improvement of innovation-based projects that are effective and productive in anticipating, diagnosing, and overseeing chronic just as mild illnesses. Advancement of the clinical artificial intelligence is distinctive to that of the nonclinical AI applications which depend on unadulterated statistical and probabilistic methodologies, as it follows representative models of infections while examining the connection between patients' indications and symptoms (Colburn, 2015; Ekman et al., 2011). These innovations are significant as they not just assist in giving treatment and diagnosis, rather they contribute toward the improvement of treatment protocol and patient's checking (Colburn, 2015; Ekman et al., 2011; Haynes et al., 2002).

Artificial Neural Networks (ANNs) are perhaps the most noticeable models in regard to the role of artificial intelligence (AI) in clinical as the greater part of AI specialists, following their motivation created ANNs. This is an endeavor to reproduce the nerve cell organizations of the brain (Barfield, 2015). This sort of approach includes the replication of the biological neural organizations for an autonomous functioning and is unique in relation to the traditional processing which needs help through human brain calculation. This wide acknowledgment of artificial intelligence, explicitly in the field of medical care, is related to the intricacies of the cutting-edge medical services industry alongside the advancement of infections which requires an in-depth data for diagnosis (Barfield, 2015; Kumar, 2017). The innovation is novel in gathering medical impediments to address various necessities, by the improvement of clinical applications that depend on computing capabilities bringing about the utilization of strategies which are valuable in helping clinicians in medical care. Artificial intelligence through its profoundly proficient outcomes is presently serving for every one of the three classical clinical undertakings which incorporates infection diagnosis, its anticipation and treatment. Basically, the clinical diagnosis cycle includes six stages. Starting from the principal period of the cycle where patient is observed, driving toward the assessment of sickness where significant information is being gathered (De Gruyter et al., 2011; Emakhu, 2019; Kuroi & Shimozuma, 2004).

6.1.2 AI technology versus COVID-19 pandemic

The spread of the worldwide pandemic COVID-19, which was first recognized in December 2019 in China, has been distinguished as the most dangerous infection brought about by the SARS-CoV-2 virus. The role of artificial intelligence is vital and fills in as most possibly amazing tool in the battle against the spread of the developing pandemic (Luengo-Oroz & Pham, 2020; Magid et al., 2021). Artificial intelligence, in such manner, can be alluded as the Natural Language Processing (NLP), Machine Learning (ML), and Computer Vision applications, which utilize big data-based models for pattern acknowledgment, its clarification alongside the forecast. These functions can be of more prominent interest in perceiving, foreseeing, and fostering a treatment plan for the COVID-19 virus infection, while dealing with its socioeconomic effect (King, Cole, & Farber, 2017; Zhang et al., 2021).

Quick and precise recognition of COVID-19 is progressively crucial to forestall the sources of infections just as assisting patients with deterring disease progression. Soft computing strategies, such as fuzzy logic, neural networks, and genetic algorithms, have demonstrated as possible apparatuses in the disease identification. They can uphold decision making, accommodating quick isolation and fitting patient treatment. Several strategies have been proposed for identifying coronavirus diseases; nonetheless, the ideal discovery precision has not been reached at this point (Udgata & Suryadevara, 2021). Fuzzy Logic (FL) depicts frameworks in terms of a mix of numeric and linguistic (emblematic). This enjoys upper hands over unadulterated mathematical (numerical) approaches or unadulterated emblematic methodologies in light of the fact that regularly framework knowledge is accessible in such a combination.

Notwithstanding, for the powerful administration of the worldwide pandemic, numerous specialists and clinical specialists have joined the utilization of artificial intelligence alongside other information instruments to battle against the common effect of the illness (McWilliams, 2020; Udgata & Suryadevara, 2021). Due to this, six regions have been recognized where artificial intelligence may contribute to successful management of the pandemic. This incorporates early admonitions and timely cautions, anticipating and following the predominance of the infection, development of information dashboards, analysis and prognosis, treatment and cure, bringing about social control. Artificial intelligence is further valuable in following, estimating, and foreseeing the spread of COVID-19 virus by the progression of time.

6.2 An introduction to deep learning

Deep learning is an AI (artificial intelligence) function that mimics the activities of the human mind in processing information and making patterns for use in making decision (Mathews, 2019). Also, deep learning is a subset of ML in artificial intelligence that has networks fit for learning unsupervised from information that is unstructured or unlabeled (Tien, 2017). Otherwise called deep neural learning or deep neural network, it can also be defined as an AI function that emulates the activities of the human mind in handling information for use in identifying objects, perceiving speech, deciphering languages, and making decisions.

Deep learning has advanced inseparably with the digital era, which has achieved a blast of information in all structures and from each part of the world (He & Bai, 2021). This information, referred to just as big data, is drawn from sources such as social media, e-commerce platforms, internet search engines, and online films. This tremendous amount of information is promptly open and can be shared using fintech applications such as cloud computing.

Notwithstanding, the information, which typically is unstructured, is huge to such an extent that it could require a very long time for people to grasp it and extract significant data (Alkatheeri et al., 2020; Constantiou & Kallinikos, 2015; Shorten & Khoshgoftaar, 2019; Thakuriah, Tilahun, & Zellner, 2017). Organizations understand the incredible potential that can come about because of unwinding this abundance of data and are progressively adjusting to artificial intelligence frameworks for automated support. Deep learning disentangles gigantic amount of unstructured information that would typically take people a very long time to comprehend and process.

Deep learning, a subset of ML, uses a progressive degree of artificial neural networks to do the cycle of machine learning (Currie, 2019; Géron, 2019; Shorten & Khoshgoftaar, 2019). The artificial neural networks are developed to be like the human brain, with neuron nodes associated together like a web. While conventional programs assemble investigation with information in a linear manner, the progressive capacity of deep learning frameworks empowers machines to handle information with a nonlinear methodology.

A specialized definition of deep learning is the utilization of neural networks with multiple or two layers. A neural organization layer is regularly made out of a parametric, nonlinear change of an input (Akçakaya et al., 2019; Courbariaux, Bengio, & David, 2015; Zhang, Wang, & Liu, 2018). Stacking these changes shapes a factual information structure equipped for planning high-dimensional contributions to yields. This planning is executed by improving the parameters. Slope descent is the device of choice for this streamlining. Slope descent functions by taking the partial derivative of the loss function with respect to each of the parameters and refreshing the parameters to limit the loss function. Deep learning gets the name "Deep" due to stacking several of such layers. The second piece of the name, "Learning," references parameter streamlining. Best in class deep learning models ordinarily contain between 100 million to 10 billion parameters and 50–200 layers (Boutaba et al., 2018; Montellano, 2020). Two of the biggest models publicly revealed are made out of 600 and 175 billion boundaries (Cavallo & Rigobon, 2016; Zeng et al., 2011). Increasing the size of these models has sped up very rapidly in the previous few years.

6.3 Literature review

6.3.1 Deep learning algorithms

Deep learning algorithms are significant and are explicitly acceptable in discovering little subtleties of the visual information that are frequently undetected by the human eye (Wurtz et al., 2019; Yannopoulos, Andronikou, & Varvarigou, 2008). In such manner, COVID-Net has been viably prepared on COVIDx, which fills in as the public information base and comprises of 16,756 chest X-beams across 13,645 patient cases from coronavirus as well as of different kinds of lung diseases. This sort of information is useful in empowering the deep learning model to isolate the qualities, which independently characterizes each kind of disease while distinguishing the new X-ray beams. As affirmed by Wong, though the model is not prepared for production, still preliminary outcomes are of essential worth in recognizing the difference between COVID-19 and other comparable infections (Sinclair et al., 2021; Soufi, Hekmatnia, & Nasrollahzadeh, 2020). In any case, the impromptu creations in the model will be held by the accessibility of extra information (Miranda & Kavan, 2005; Shankar et al., 2020; Solima, Della Peruta, & Maggioni, 2016).

Wong also added that addition of the enormous sample size would have a gigantic effect in improving COVID-Net which is utilized to foster new learning models to identify COVID-19 infections. It is additionally recommended that chest X-rays and other CT scans are as yet extensive as correlative instruments and can be utilized in districts encountering the short stock of the testing units. Furthermore, there are sure circumstances which require CT scans and different chest X-rays for the positive analysis of the viral tests to recognize the power of spread of illness for additional treatment and care arranging. Wong underscored on the productivity of deep learning innovation in decreasing the burden on radiologists by empowering various frontline healthcare workers with least expertise for better analysis (Henriksen, 2013; Makarius et al., 2020; Matteo, 2012).

6.3.2 Machine learning applications

Generally, computer algorithms utilized for artificial intelligence explicitly depend on Machine Learning (ML) methods which, from a more extensive perspective, incorporate computer vision and natural (Al-Sahaf et al., 2019; Dunjko & Briegel, 2017). Wiens and Shenoy (2018) directed a new audit on the convenience of ML in medical care epidemiology and characterized it as the investigation of strategies and instruments to recognize patterns in information (Zekiria & Bachelor, 2018). ML methods are created by using a bunch of algorithms, for example decision trees, logistic regression, or deep learning which can be classified into unsupervised, supervised, and reinforcement learning procedures. Contributions of every class are different, as unsupervised learning toward one side gives strategies to information clustering, while supervised learning is focused toward illness classification. Roth et al. (2020) gave a detailed examination in regard to the value of ML in medical services sector, where patients' clinical records have been generally put away as electronic healthcare records at various worldwide medical services establishments (Laker et al., 2018). Contributions of ML in medical services industry are viable as chronic conditions like sepsis require a prompt determination alongside the therapeutic activities, explicitly when causative pathogens remain unidentified (Benjamin et al., 2018; Dellinger et al., 2008).

Gigantic molecular testing for COVID-19 virus has been pointed as crucial to direct the spread of the illness. Pooling techniques can upgrade testing proficiency, yet they are practical just at low prevalence of the illness. Thus Smart Pooling has been proposed by Leslie et al. (2020), a machine learning technique that utilizes sociodemographic information from patients to expand the proficiency of pooled molecular testing for COVID-19 virus by organizing tests into every negative pool. Efficiency gains of 42% regarding singular testing at sickness pervasiveness of up to 25% were shown, a system in which two-step pooling offers negligible effectiveness gains.

Pandey et al. (2020) directed an investigation to outline the methodological methodologies and prerequisites for the ideal utilization of ML in future disease management. A lifelong learning application was created in this investigation, to convey the exact data with respect to the virus (Madhavan et al., 2021; Solanki et al., 2020; Tande & Patel, 2014). Significant functions are performed through the coordinating sources of bona fide and checked data extracted from the reports, for example, those proposed by the World Health Organization (WHO) through natural language synthesis and ML. The application is helpful as it uses the extraordinary text to foster distinctive speech engines. The methodology was further valuable in giving helpful data in regard to sterilization, hygiene, and water which is essential to the current advancement of the developing pandemic.

6.3.3 Robotics applications

Worldwide economic and public activity has been seriously tested since the World Health Organization (WHO) pronounced the COVID-19 sickness a pandemic (Antipova, 2021; Litewka & Heitman, 2020). Travel, hospitality, and tourism, specifically, have been enormously affected by the lockdowns used to keep up social distance to deal with the sickness. Robotics, AI, and human-robot interactions have acquired an expanded presence to help deal with the spread of COVID-19 virus in airports, hospitals, transportation frameworks, amusement and grand regions, hotels, restaurants, and communities in general (Agarwal et al., 2021; Van et al., 2020). Humanoid robots, self-driving vehicles, drones, and other intelligent robots are utilized from multiple points of view to diminish human contact and the likely spread of the SARS-CoV-2 infection, including conveying materials, cleaning and disinfecting public spaces, recognizing or estimating body temperature, giving safety or security, and consoling and entertaining patients. While disputable in the past because of worries over work losses and information security, the adoption of robotics and AI in tourism, travel, and health industry will probably proceed after the COVID-19 pandemic is less serious.

Artificial intelligence and its competent features when conveyed in medical services area call for profoundly assorted applications, alongside the advancement of various groups with long-term associations (Sharma, Mahapatra, & Agarwal, 2020; Singla, 2020). Robotics is another such sector where artificial intelligence innovation when utilized aids in battling against COVID-19 pandemic. These robots are utilized for cleaning, sanitizing, and logistics purposes that require human contribution. Mullane and Bischofberger (2020) in their new article in *Science Robotics* illustrated that robots can be a viable asset in battling the pandemic, as their administrations can be used for medication and food conveyance, cleaning, alongside gathering data in regard to the most critical signs.

The robots have a friendly face, explicitly intended to keep patients at ease. To secure doctors and medical caretakers, these advanced mechanics help in decreasing the need of face masks and personal protective equipment (PPE), explicitly at the hour of their short inventory in these emergency healthcare facilities. These robots further empower patients to speak with the clinical sensors, as they are created by using more than 60 sensors giving them the ability to make significant voice collaborations, voice confinement, face acknowledgment, and video chat, by keeping away from impediments through auto charging feature (Sharma, 2021; Wang & Parsons, 2017). In the situations where the greater part of the manufacturers has confined themselves into their homes to keep away from contact with infection, numerous manufacturers have centered in changing their production lines to the general automation (Ashima, Haleem, & Bahl, 2021; Elavarasan, 2020). In spite of the fact that contributions of robots in the field of medical services are irrefutable, nonetheless the developing pattern of the robotics is as of now making a tremendous effect over low-income jobs in the developing nations.

For example, Nanyang Innovative College has created XDBot, a sanitization robot controlled with a computer or tablet, see Fig. 6.1 (AAAS, 2020). XDBot utilizes an electrostatically charged nozzle to splash positively charged synthetics onto negatively charged surfaces for exceptionally compelling cleaning.

Moreover, XDBot has a 6-axis arm, which permits it to have a more noteworthy reach contrasted to other robots (Holland, Kingston, & McCarthy, 2021). This empowers it to spray regions thought to be inaccessible, like light switches and door handles. Adding this with an 8.5-L tank and 4-h run time, XDBot can exceed human cleaning limit, which makes it a valuable device for disinfecting indoor public regions (Holland et al., 2021). Efforts have additionally been made to utilize comparative spray advancements to clean bigger regions in a more limited measure of time. These incorporate tank-style robots that can climb stairs substantially more effectively than wheeled robots with comparative innovations. These robots are additionally controlled by remote controls, making them simple to operate.

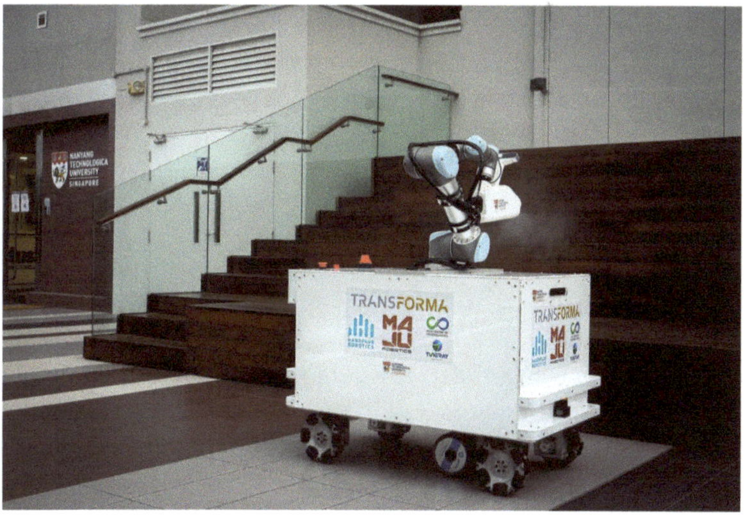

Fig. 6.1 XDBot. *(Courtesy AAAS. (2020).* NTU Singapore researchers build disinfection robot to aid cleaners in COVID-19 outbreak. *EurekAlert! Science News. Available at: https://www.eurekalert.org/pub_releases/2020-04/ntu-nsr041420.php. (Accessed 21 June 2021).)*

Fig. 6.2 ZhenRobotics Corp hand cleaning robot.

While a few robots have been intended to vacuum floors and others to spray sanitizer, ZhenRobotics Corp has developed a versatile hand cleaning robot, see Fig. 6.2 (ZhenRobotics, 2020). This robot can naturally explore through external regions and convey hand sanitizer to people. Robots like these will be extremely helpful when attempting to guarantee a high number of people in a populace approach sanitary items, possibly decreasing the spread of COVID-19 virus.

6.4 Materials and methods

6.4.1 Determination of the force exerted by the swab on the nasal passage

The nose of human beings differs depending on racial and ethnic groups in terms of anatomy and morphology. Therefore it is not ideal to assume a standard size of the nose and nostril location. The human nose consists of two nasal valves, the external and internal nasal valves. The nasal valve is responsible for air resistance inside the nasal cavity. This suffice to say, resistance force will increase as the swab is forced inside the nostrils. So, the standard force that must be applied while the swab is forced inside the nostrils will differ starting from nostril entrance until sample collection. The maximum

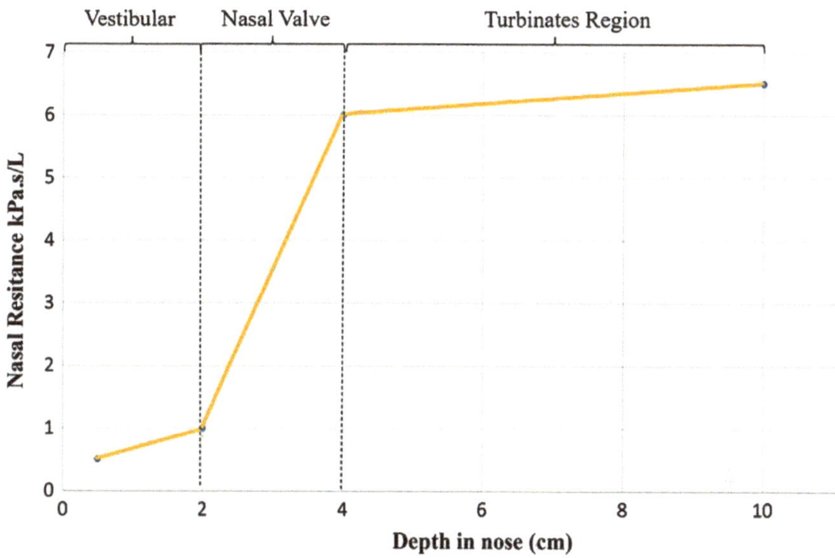

Fig. 6.3 Nasal resistance against depth in nose.

force that the swab can apply to the nasal passage walls is assumed to be 1 N. Note that this force has been adapted from previous researches that were done for surgical robots and hence to obtain the accurate values a research can be done but this is beyond the project's scope. The main focus of this project is to determine an algorithm that can be used to collect the sample through the nostrils and hence the other factors are extrinsic. After that, there is the determination of the distance between nostril entrance and ear opening. This is the distance the swab has to move inside the nostrils during sample collection. To obtain this, the following graph has been used (see Fig. 6.3).

6.4.2 Determination of the nostrils position

From the earlier discussion it can be seen that different people have diverse noses and hence different nostrils position. Therefore when determining nostrils position the system has to adapt each and every time a new patient is being tested. Thus a facial recognition system has been adapted. The facial recognition system is of paramount importance because it is used first during patient registration and for nostrils identification. During registration when patients are filling in their personal details in an LCD located at booth front, a camera will be used to capture their faces at different angles. Thus the facial recognition will be used to permit a registered user with the same face to enter the booth, else they have to redo the registration process (see also the following sections). Assuming the patient is now seated on a chair inside the booth at a known distance from the robotic end effector. A camera has been attached to the end effector, which will be used to detect the face of someone seated at the chair. This chair has been designed such that the patient's head will be able to be tilted at an angle suitable for easy sample collection. The facial recognition attribute will be used to number most important parts of the face and in this report a pretrained facial landmark detector integrated within python dlib-library has been used to approximate the location of 68 (x, y) coordinates that map to facial structures on the face (see the following figure). The most intrinsic parts will be the nose and mouth, as these are where the sample will be collected. In this report, the priority will be to collect sample in the nostrils. So, the following procedure is used for sample collection in the nostrils.

From Fig. 6.4, it can be shown that facial landmarks numbering for the nose start from 28 to 31 for the septum and 32 to 36 for nostrils.

Comparing the two figures earlier (see Fig. 6.5), it can be shown that the nostrils are located between 32 and 36 for original resolution and 32 and 36 for higher resolution but the outer points seem to have shifted closer, that is, when the camera is closer to the patient's face. Therefore the end effector camera resolution is continuously changing as the camera moves toward the patient's face and so does the landmark coordinates. To obtain precise location of the nostrils, original resolution can be compared with the other resolutions. So, to maintain original resolution, another camera has been fixed at a position such that it always captures the initial resolution. Now, in order to identify the nose, two coordinates will be compared and these are as follows:

✓ coordinates from the camera not fixed on the arm and
✓ novel coordinates from the camera attached to the robotic manipulator's end effector.

After that, there is determination of the nostril opening and for the first part, nostril opening coordinates are determined as the midpoints between points 32, 33 and 35, 36. So given the coordinates of these points as (x_{32}, y_{32}); (x_{33}, y_{33}); (x_{35}, y_{35}); and (x_{36}, y_{36}), then nostrils opening A and B coordinates are given as follows:

$$C_A = \left(\frac{x_{32} + x_{33}}{2}, \frac{y_{32} + y_{33}}{2}\right)$$

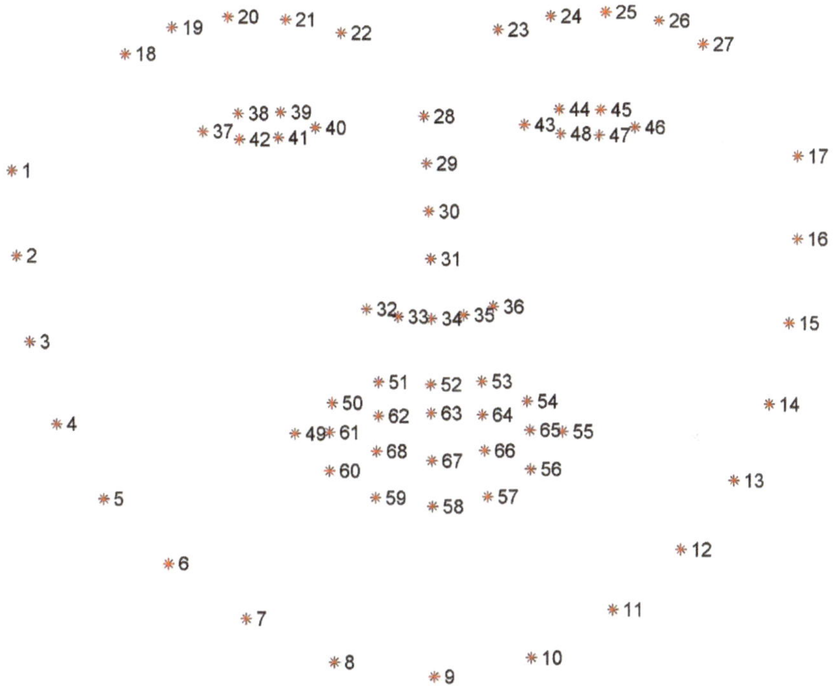

Fig. 6.4 Sixty-eight facial landmark coordinates from the iBUG 300-W data set (higher resolution). *(Courtesy Rosebrock, (2017). Deep learning for computer vision with python, PYIMAGESEARCH. https://www.pyimagesearch.com/deep-learning-computer-visionpython-book/.)*

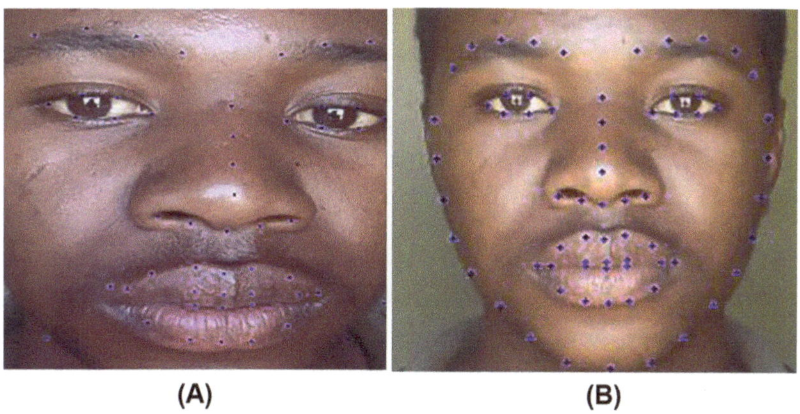

Fig. 6.5 Landmarks on a high resolution (A) and original smaller resolution (B). *(Courtesy Marvellous Moyo and Tawanda Mushiri.)*

$$C_B = \left(\frac{x_{35} + x_{36}}{2}, \frac{y_{35} + y_{36}}{2}\right)$$

For the second part, nostrils openings coordinate A and B are assumed to be located at points 33 and 35, respectively, due to the increased resolution. To identify the exact location of nostril openings, the two outputs will be compared. If the deviation between the two outputs is small then the first part will be executed. If not, the results will be compared with force sensor readings when the swab first touches the nostrils at entrance. This else part involves first checking coordinates of second part against force sensor results. If the force is very small, then it will be executed, else first part results will be compared with force sensor readings and then executed. The assumption is that, first part should be executed when compared with force sensor readings since it is computed as a midpoint and hence the probability of locating the nostril opening should be high (see Fig. 6.6). For very advanced system beyond the scope of this project, facial expression of the patient can also be considered during location of the nostril opening.

6.4.3 Determining how to align the swab with nostril opening

After the location of nostril openings, the swab has to be in alignment with an opening, otherwise during movement inside the nostrils it would collide on nasal passage surfaces instead of going straight in. To prevent this from happening the following method will be used. First, the facial recognition system is used to locate landmarks at patient's mouth (see earlier figure). From the earlier figures, points 51, 52, and 53 will be used. After that, the end effector will align the swab with nostril point 34 such that they will be collinear in the horizontal plane (x–z plane), that is, perpendicular to the face plane (x–y plane). The facial recognition landmark attributes give two coordinates (x, y) and hence it is a 2D system. Therefore points 34 and 52 lie on the same plane and the vertical distance between them can be calculated as follows:

$$l_2 = y_{34} - y_{52}$$

Now an ultrasonic sensor is used to measure the distance between end effector and points 34 and 52. These distances are given as l_{34} and l_{52}, respectively. Therefore the difference between these two lengths gives (see Fig. 6.7)

$$l_1 = l_{34} - l_{52}$$

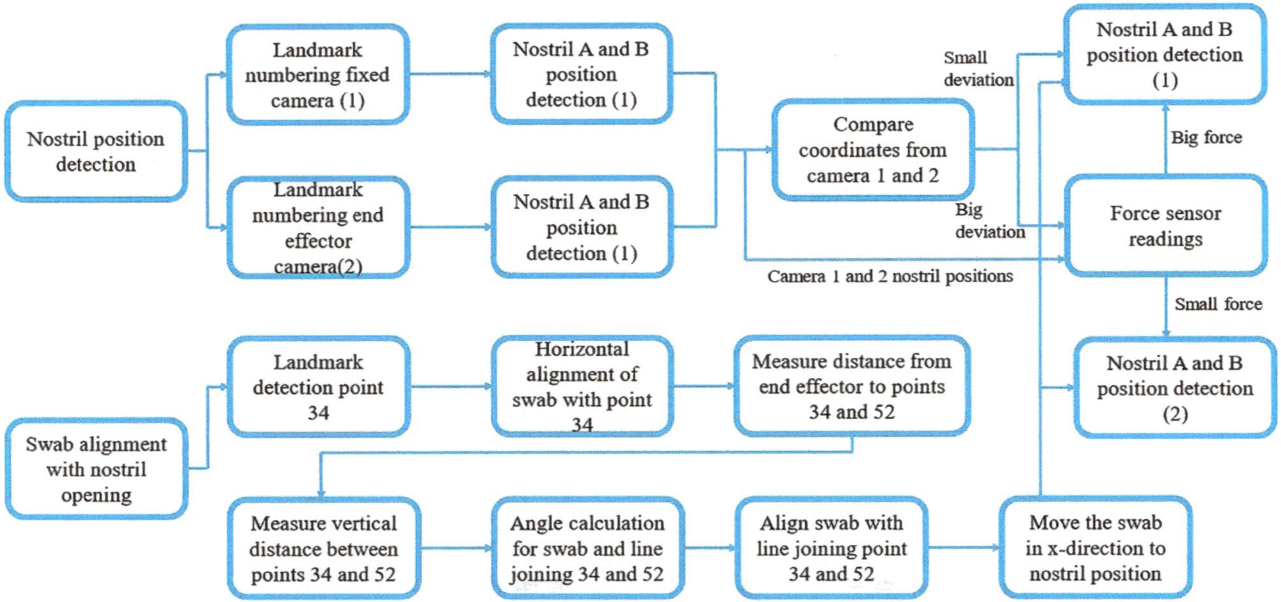

Fig. 6.6 Swab alignment and nostril position detection. *(Courtesy Marvellous Moyo and Tawanda Mushiri.)*

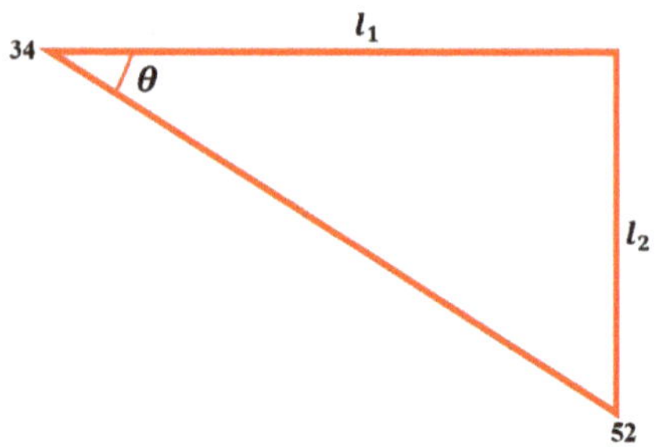

Fig. 6.7 End view of line joining nose and mouth with horizontal swab. *(Courtesy Marvellous Moyo and Tawanda Mushiri.)*

Now the angle between the swab which is collinear with point 34 and the line joining points 34 and 52 is θ obtained by using tangent (see the earlier figure). Therefore, in order to perfectly align the swab with the nostril opening, its angular position has to be adjusted twice. These two steps are as follows:

1. align the swab with nostril point 34
2. align the swab with line joint points 34 and 52 at an angle θ to point 34.

Therefore the swab has to align with nostril opening first then locate the opening, but point 34 is not at the opening. So, after alignment the swab has to move from point 34 to the opening point while maintaining its alignment. This can be achieved by moving the swab in the x-direction while keeping the y and z values constant. Note some angular positions of the swab will be restricted such that their occurrence is minimal compared to others. Still, they have been included so that in the event joints align the swab to those positions, execution of such motions occurs.

6.4.4 The testing process of the intelligent robotic testing machine

The testing process will be explained starting from person's entrance until they exit the booth. A mobile phone or iPad will be provided to patients at the waiting area. The phone will be connected to the internet and will display a web application, consisting mainly of a registration form which must be filled by every patient who enters the booth to be tested (note should be taken to improve the rate of testing patients can fill the form using their mobile phones and then go for testing). The form will ask patients to fill in their personal details including their medical conditions. The key thing is

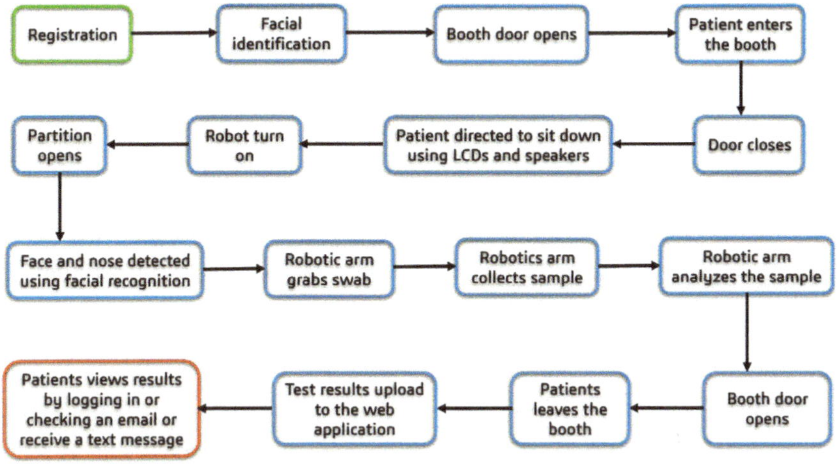

Fig. 6.8 The testing process layout of the intelligent robotic system. *(Courtesy Marvellous Moyo and Tawanda Mushiri.)*

to upload a profile picture during registration. At the booth door a webcam has been attached and the profile picture will be used to unlock the door using a facial identification software. Fig. 6.8 summarizes the whole process during testing.

After entrance inside booth, LCDs and loud speakers will be used to guide the patient on what they should do, for example asking one to take a seat or briefing them on how the testing will be conducted, the type of tests that will be done. Thereafter the septum separating the patient and testing robotic manipulator will be opened (refer with the following chapter). The robotic manipulator will then grab a swab on the table and it will approach the patient's nose and collect the sample. The person's nose and face will be detected using the facial recognition landmarks system. An infrared camera will be used to determine the temperature variations on one's body (mainly the face). After collecting the sample, the testing will be done depending on the type of test being conducted. Since the testing process is done automatically by the robot and there is no human supervision, an intelligent control system has to be adapted. In this report the research has used the fuzzy logic control system to better manage the testing process. The main parts to be controlled in this system are the door, partition, flow rate of the air compressor, and the robotic manipulator motion during the testing process. The design of this control system is explained following, starting with the robotic manipulator control during the testing process.

6.4.5 Robotic manipulator control during sample collection

The first step for collection of the sample is the identification of the nose and nostrils. To achieve this, a facial recognition software has been adapted together with the data obtained from human nose study (see previous sections). This robotic manipulator is of paramount importance in the testing process because it collects samples, mixes them with chemicals, and then does the testing and hence its control system must be designed such that it performs these tasks with minimum errors as possible. Thus fuzzy logic has been employed. The basis of the fuzzy logic control is to know what needs to be controlled, how it will be controlled, and what are the outputs of a system. This can be visualized in Fig. 6.9.

In this system, the variables that need to be controlled are the distance between the patient's nose and end effector, angular position of the swab with respect to a nostril, force exerted by the swab on nasal passage, and lastly the distance the swab has to move inside the nostrils until sample collection. These variables are the input to the fuzzy logic controller and the output will be the velocity of the end effector. For one to be able to clearly define the input variables, a deep understating of the human nasal passage must be known (the human respiratory system). Therefore some inputs have been designed based on findings obtained from the scientific study of the human nose.

6.5 Results

6.5.1 Proposed web application

For the simplicity of the design and representation purposes, the registration from will consist mainly of the following: forename, surname, date of birth, national identity number, email address, phone number, profile picture upload section, and password. Fig. 6.10 shows the proposed registration form.

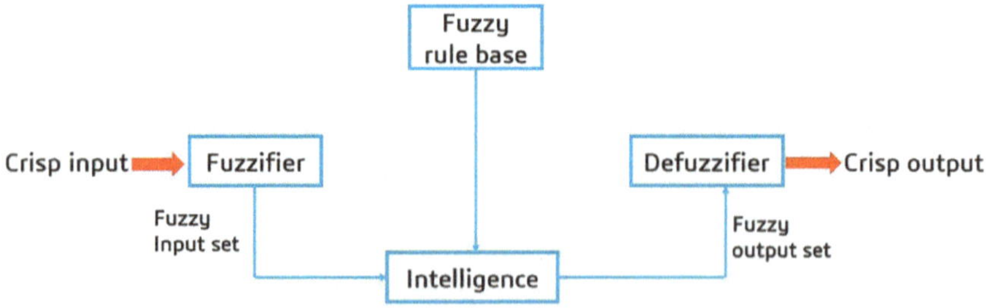

Fig. 6.9 The fuzzy controller architecture. *(Courtesy Marvellous Moyo and Tawanda Mushiri.)*

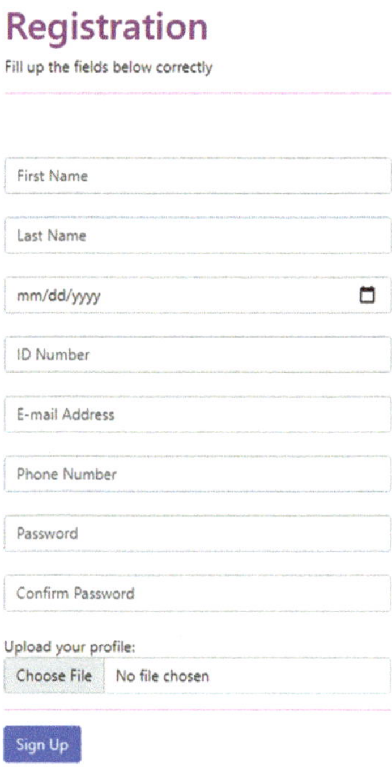

Fig. 6.10 The proposed registration forms. *(Courtesy Marvellous Moyo and Tawanda Mushiri.)*

When the patients open the web application, they will be taken first to the login page after that they follow the instructions outlined on that page. For the registered patients they can login their details to see their test results anywhere and anytime of the day without the need to visit the hospital. Also, at the same time results will be sent to them via email and text messages to their mobile phones. The registration form and the welcome page are shown in Figs. 6.11 and 6.12.

6.5.2 Design of the fuzzy logic controller

After defining the input and output variables (see Chapter 3), then one has to identify corresponding linguistic inputs and outputs. The input variables to the fuzzy logic controller will be provided by the sensors. The fuzzy sets for the input variables are outlined here. First is the distance between the nose and end effector (D_{NE}), which will be measured using an ultrasonic sensor. The fuzzy sets for this input are Very Far (VF), Far (F), Close(C), and Very Close (VC). Next is the angular position of the swab with respect to a nostril (θ_{SN}) which will be determined using the Cartesian coordinates of the

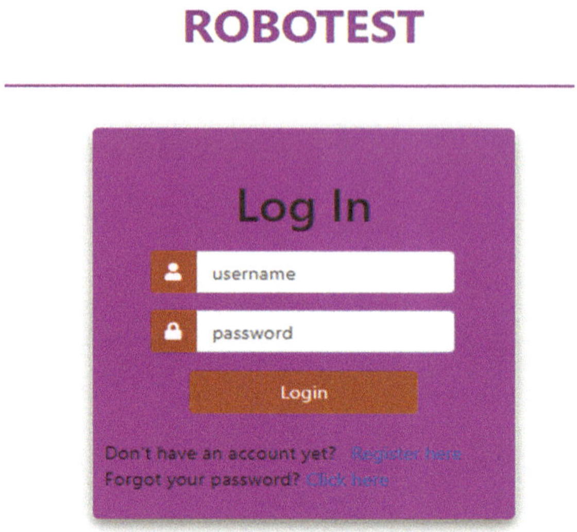

Fig. 6.11 Log in page. *(Courtesy Marvellous Moyo and Tawanda Mushiri.)*

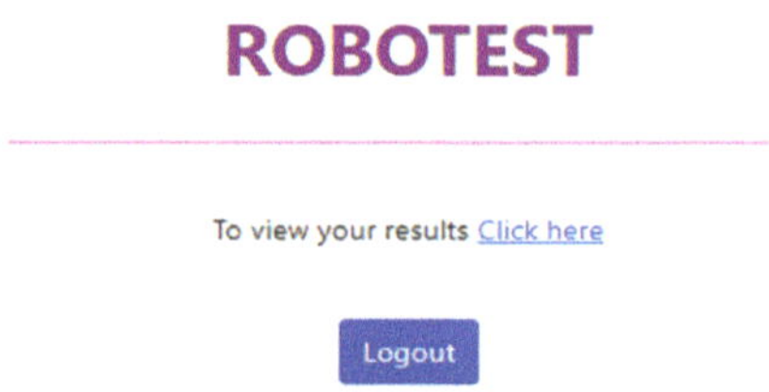

Fig. 6.12 Welcome page. *(Courtesy Marvellous Moyo and Tawanda Mushiri.)*

manipulator and corresponding fuzzy sets are Negative (N), Small Negative (SN), Zero (Z), Small Positive (SP), and Positive (P). After that, is the swab distance moved inside nostrils (D_{IN}) and this will also be measured using an ultrasonic sensor. The fuzzy sets are given as at entrance (A), Just (J), Half Way (HW), and Way Inside (WI). Lastly is the force exerted by the swab on the nasal passage surfaces (F_{NP}) and this will be measured using a force sensor attached at the end effector and the fuzzy sets are given as Very Small (VS), Small (S), Standard Size (SS), Large (L), and Very Large (VL). The membership functions for these inputs are shown in Figs. 6.13–6.16.

These inputs are then subjected to a fuzzy if-then rule base that determines objectives for each behavior. After this, the inputs are defuzzified to convert them to outputs. Two defuzzified outputs have been defined and these are end effector velocity and spherical wrist angular position.

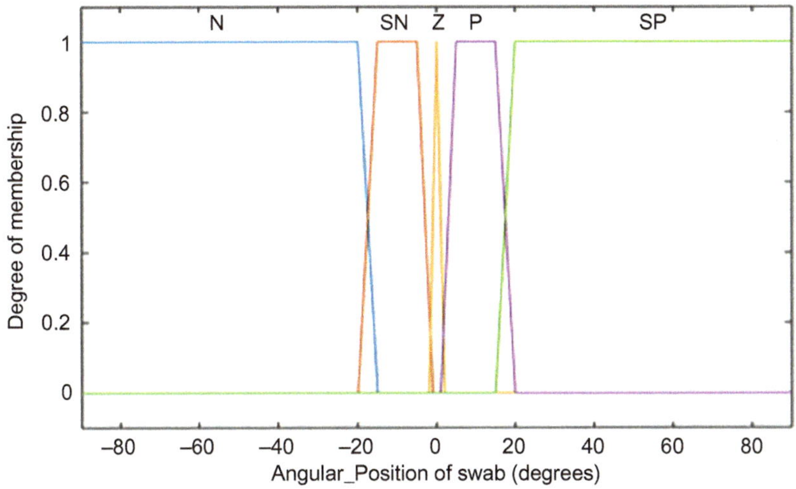

Fig. 6.13 Angular position of swab (degrees). *(Courtesy Marvellous Moyo and Tawanda Mushiri.)*

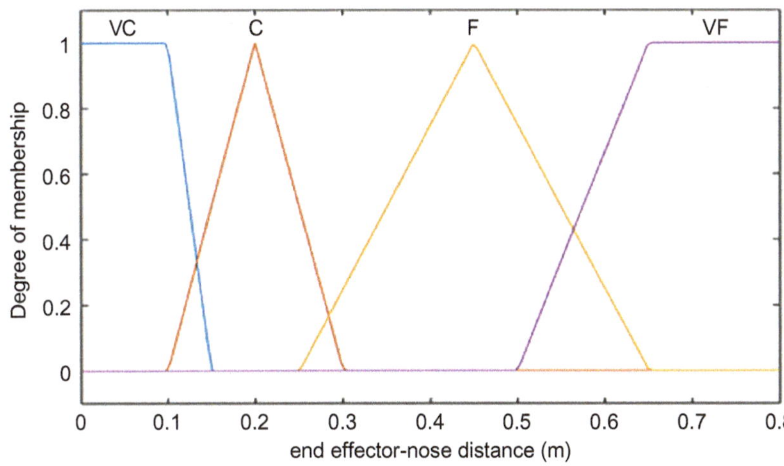

Fig. 6.14 End effector-nose distance. *(Courtesy Marvellous Moyo and Tawanda Mushiri.)*

Angular position of the wrist will be controlled mainly by varying angular velocity of link 4. The fuzzy sets for end effector velocity are Zero (Z), Small Positive (SP), and Positive (P). For the angular position of the spherical wrist, fuzzy sets are given as Much Down (MD), Down (D), Remain (R), Up (U), and Much Up (MU). Figs. 6.17 and 6.18 show the membership functions for these outputs.

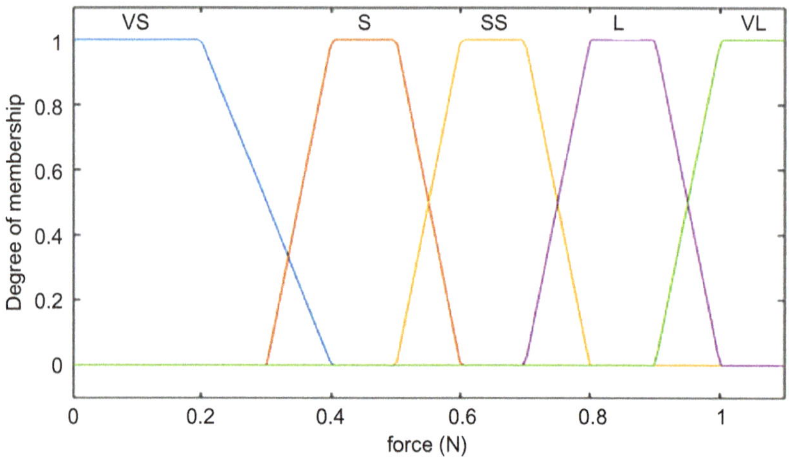

Fig. 6.15 Force exerted on the nasal passage surfaces. *(Courtesy Marvellous Moyo and Tawanda Mushiri.)*

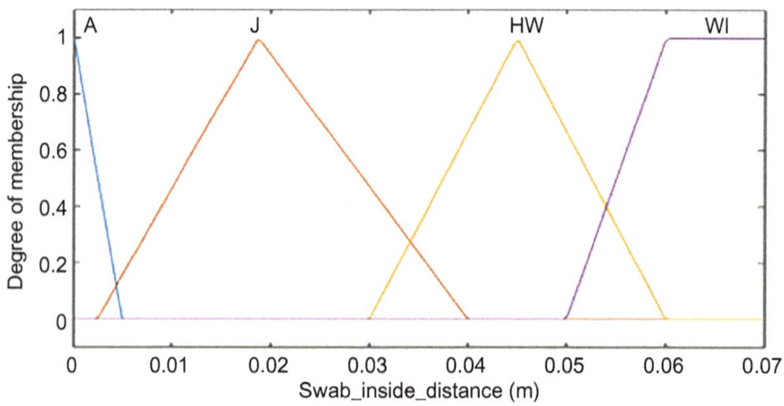

Fig. 6.16 Swab inside distance. *(Courtesy Marvellous Moyo and Tawanda Mushiri.)*

The fuzzy rule base for this system has been divided into two parts (see Tables 6.1–6.3). The first part compares angular position of swab with respect to a nostril and end effector to nose distance. The second part will compare distance moved by swab inside the nose and force it applies on nasal passage. The following examples are used to show how the rules are computed for each input behavior.

✓ If D_{NE} is VF and θ_{SN} is N THEN end effector velocity is P
✓ If D_{NE} is VF and θ_{SN} is N THEN Angular position of wrist is R
✓ If D_{IN} is J and F_{NP} is VS THEN end effector velocity is P
✓ If D_{IN} is J and F_{NP} is VS THEN Angular position of wrist is R

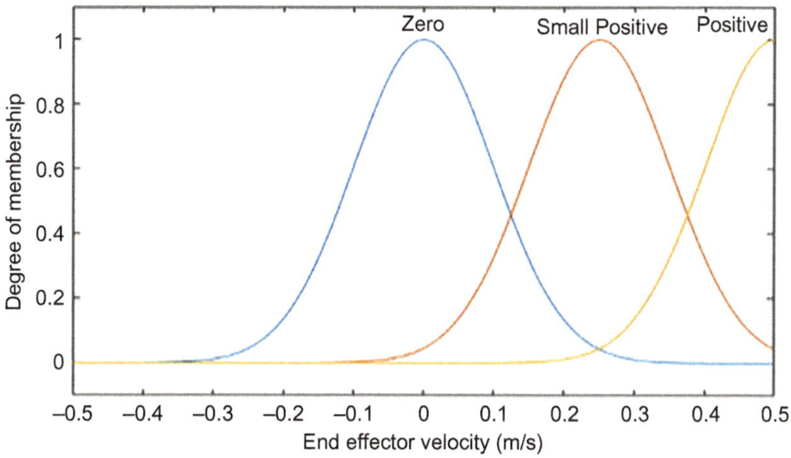

Fig. 6.17 End effector velocity. *(Courtesy Marvellous Moyo and Tawanda Mushiri.)*

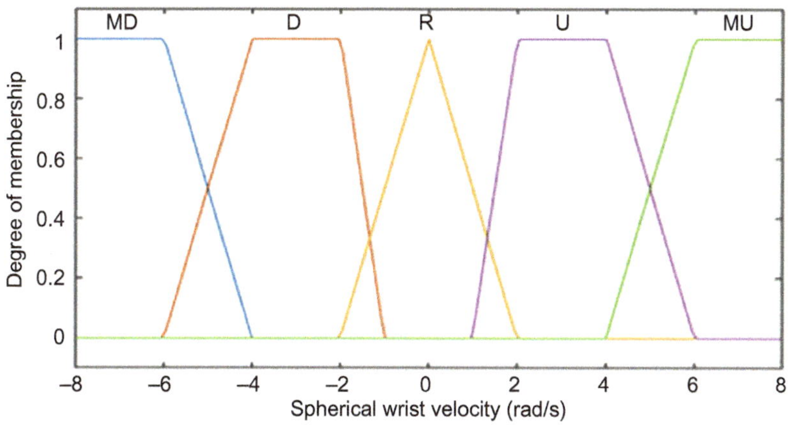

Fig. 6.18 Spherical wrist angular velocity. *(Courtesy Marvellous Moyo and Tawanda Mushiri.)*

6.6 Discussion

Artificial intelligence, robotics, and advanced technologies have given significant help during the COVID-19 pandemic and shown their promising future in the public eye. Nonetheless, challenges stay for these technologies to be applied broadly and incorporated into a drawn-out shift in way of life.

Cost is the greatest obstacle for these technologies, especially robotics, to be carried out broadly. Despite the fact that robots have extraordinary potential as apparatuses to address different issues in clinical and everyday

Table 6.1 Fuzzy logic rule base part one.

	If			Then	
Rule no.	D_{NE}		θ_{SN}	End effector velocity	Angular position of wrist
1	VF	and	N	P	R
2	F	and	N	P	R
3	C	and	N	SP	R
4	VC	and	N	Z	MU
5	VF	and	SN	P	R
6	F	and	SN	P	R
7	C	and	SN	SP	R
8	VC	and	SN	Z	U
9	VF	and	Z	P	R
10	F	and	Z	P	R
11	C	and	Z	SP	R
12	VC	and	Z	Z	R
13	VF	and	SP	P	R
14	F	and	SP	P	R
15	C	and	SP	SP	R
16	VC	and	SP	Z	D
17	VF	and	P	P	R
18	F	and	P	P	R
19	C	and	P	SP	R
20	VC	and	P	Z	MD

life, the significant expense of robots implies they will in all probability be available just to the richest emergency clinics and urban communities, where just a little part of individuals can profit. Another worry with robots is security. Numerous conveyance robots have been carried out in medical clinics and public regions during this pandemic to lessen individual-to-individual contact, and they can be explored to their objections consequently with artificial intelligence or different technologies. Notwithstanding, the safety of these conveyance robots is as yet problematic, particularly in clinics. Considering the little passageway room in medical clinics, any increment of individuals and transportation in medical clinics will create further problems for the robot navigation.

Security and confidentiality are another worry, since certain nations may follow the spread of COVID-19 by checking individuals' every day exercises through artificial intelligence-driven facial recognition or smartphone

Table 6.2 Fuzzy logic rule base part two-one.

	If			Then	
Rule no.	D_{IN}		F_{NP}	End effector velocity	Angular position of wrist
1	A	and	VS	SP	R
2	J	and	VS	P	R
3	HW	and	VS	P	R
4	WI	and	VS	SP	R
5	A	and	S	SP	R
6	J	and	S	P	R
7	HW	and	S	P	R
8	WI	and	S	SP	R
9	A	and	SS	Z	Section 1 in Table 6.3
10	J	and	SS	P	R
11	HW	and	SS	SP	R
12	WI	and	SS	SP	R
13	A	and	L	Z	Section 2 in Table 6.3
14	J	and	L	–	–
15	HW	and	L	SP	R
16	WI	and	L	Z	R
17	A	and	VL	Z	Section 3 in Table 6.3
18	J	and	VL	–	–
19	HW	and	VL	–	–
20	WI	and	VL	Z	R

Table 6.3 Fuzzy logic rule base part two-two.

		If					Then
Section	Rule no.	D_{IN}		θ_{SN}		F_{NP}	Angular position of wrist
1	1	A	and	N	and	SS	MU
	2	A	and	SN	and	SS	U
	3	A	and	Z	and	SS	–
	4	A	and	SP	and	SS	D
	5	A	and	P	and	SS	MD
2	6	A	and	N	and	L	MU
	7	A	and	SN	and	L	U
	8	A	and	Z	and	L	–
	9	A	and	SP	and	L	D
	10	A	and	P	and	L	MD
3	11	A	and	N	and	VL	MU
	12	A	and	SN	and	VL	U
	13	A	and	Z	and	VL	–
	14	A	and	SP	and	VL	D
	15	A	and	P	and	VL	MD

applications. This information ought to be utilized under severe supervision because of the private data included. Notwithstanding the management of the public authority, different techniques ought to be applied to secure privacy, for example, limiting the information use period and implementing information assurance rights, including the option to erase information and the utilization strategies for the information.

A few exercises can be gained from the COVID-19 pandemic. The majority of the innovations utilized in the pandemic have been adjusted from prior technologies. In spite of the fact that exploration research facilities and new companies are making new innovations and models during the pandemic to help medical services workers and general society, their items may not arrive at the market on schedule or they might not be able to develop their client base quick enough for a critical mass. Generally, it is more proficient and versatile to repurpose existing technologies than create new ones during a pandemic or emergency.

With all this in mind, the researchers focused on using AI, deep learning, and robotics in collecting COVID-19 samples. It was realized during literature survey that most applications of robots during this COVID-19 pandemic were focused on disinfection and delivery while AI techniques such as deep and machine learning were used for screening potential cases. With the increase in number of deaths due to COVID-19, especially for the healthcare workers, there is a need to find ways in which AI can be adapted to help reduce the number of deaths. Using facial recognition applications, it has been shown that it will be possible to collect samples from patients without any human supervision. To better improve safety of patients during sample collection, fuzzy logic control has been employed. In addition, a web application has been used to collect personal information of patients without them interacting with healthcare workers. Therefore, by using such a system, not only will the testing rate of COVID-19 virus be improved, but also risks of healthcare workers' lives and costs of PPE will be reduced effectively.

6.7 Programming examples

6.7.1 Web application

Programming codes that were used to develop the web application are outlined as follows: Registration form was created using php, html, css, and JavaScript. The following code can be saved as registration.php and was tested using Xampp software.

```
<?php
 require_once('config.php');
?>

<!DOCTYPE html>
<html>
<head>
 <title>Patient Registration | PHP</title>
 <link rel="stylesheet" type="text/css" href="css/bootstrap.min.
css">
</head>
<body>
 <div>
  <form action="registration.php" method="post">
   <div class="container">

    <div class="row">
     <div class=col-sm-4>
      <h1>Registration</h1>
      <p>Fill up the fields below correctly</p>
      <hr class="mb-3">

      <label for="firstname"><b>First Name</b></label>
      <input class="form-control" id="firstname" type="text"
name="firstname" required>

      <label for="lastname"><b>Last Name</b></label>
      <input class="form-control" id="lastname" type="text"
name="lastname" required>

      <label for="dateofbirth" ><b>Date of birth<b></label>
      <input class="form-control" id="dateofbirth" type="date"
name="dateofbirth" required>

      <label for="nationalid"><b>ID Number</b></label>
      <input class="form-control" id="nationalid" type="text"
name="nationalid" required>
      <label for="email"><b>E-mail Address</b></label>
      <input class="form-control" id="email" type="email"
name="email" required>
      <label for="phonenumber"><b>Phone Number</b></label>
      <input class="form-control" id="phonenumber" type="text"
name="phonenumber" required>
      <label for="password"><b>Password</b></label>
      <input class="form-control" id="password" type="password"
name="password" required>
```

```
        <label for="profile">Upload your profile:</label>
        <input class="form-control" id="profile" type="file"
name="profile" accept="image/*" required>
        <hr class="mb-3">

        <input class="btn btn-primary" type="submit" id="register"
name="create" value="Sign Up">
       </div>
      </div>
     </div>
    </form>
   </div>
   <script src="https://ajax.googleapis.com/ajax/libs/jquery/3.3.1/
jquery.min.js"></script>
   <script src="https://cdn.jsdelivr.net/npm/sweetalert2@8"></
script>
   <script type="text/javascript">

    $(function(){
     $('#register').click(function(e){

      var valid = this.form.checkValidity();

      if (valid) {

       var firstname  = $('#firstname').val();
       var lastname  = $('#lastname').val();
       var dateofbirth = $('#dateofbirth').val();
       var nationalid  = $('#nationalid').val();
       var email   = $('#email').val();
       var phonenumber  = $('#phonenumber').val();
       var password  = $('#password').val();
       var profile  = $('#profile').val();

       e.preventDefault();

       $.ajax({
        type: 'POST',
        url: 'process.php',
        data: {firstname: firstname,lastname: lastname,dateof-
birth: dateofbirth,nationalid: nationalid,email: email,phonenum-
ber: phonenumber,password: password, profile: profile},
        success: function(data){
            Swal.fire({
             'title': 'Successful',
             'text' : data,
```

```
                'type' : 'success'
              })
              if ($.trim(data) == "sucessfully saved") {
                setTimeout('window.location.href    =    "login.php"',
500);
              }
          },
          error: function(data){
              Swal.fire({
                'title': 'Errors',
                'text' : 'There were errors while saving the data',
                'type' : 'error'
              })
          }
        });
      } else {
      }
    });
  });
  </script>

</body>
</html>
```

The following code can be named `config.php` and is used for database connection:

```
<?php
 $db_user = "root";
 $db_pass = "";
 $db_name = "covid_patients";

 $mysqli = new mysqli('localhost', $db_user, $db_pass, $db_name);
?>
```

6.7.2 Facial recognition system

Python and its libraries were used to develop the facial recognition software and the code is shown as follows:

This code can be saved as facial_recognition.py and run in the command prompt for example.

```
import cv2
from numpy import*
import dlib
```

```
cap = cv2.VideoCapture(0)
detector = dlib.get_frontal_face_detector()
predictor = dlib.shape_predictor("shape_predictor_68_face_land-
marks.dat")

def mid(v1, v2):
 result = int( (v1 + v2)/2 )
 return result

while True:
 _, frame = cap.read()
 gray = cv2.cvtColor(frame, cv2.COLOR_BGR2GRAY)

 faces = detector(gray)
 for face in faces:
  x1 = face.left()
  y1 = face.top()
  x2 = face.right()
  y2 = face.bottom()
  #cv2.rectangle(frame,(x1, y1), (x2, y2), (0, 255, 0), 3)

  landmarks = predictor(gray, face)

  for n in range(31, 36):
   x = landmarks.part(n).x
   y = landmarks.part(n).y
   cv2.circle(frame, (x, y), 1,(255, 0, 0), -1)

  x_32 = landmarks.part(32).x
  y_32 = landmarks.part(32).y
  x_33 = landmarks.part(33).x
  y_33 = landmarks.part(33).y
  x_35 = landmarks.part(35).x
  y_35 = landmarks.part(35).y
  x_36 = landmarks.part(36).x
  y_36 = landmarks.part(36).y

  x_m1 =mid(x_32, x_33)
  y_m1 =mid(y_32, y_33)
  x_m2 =mid(x_35, x_36)
  y_m2 =mid(y_35, y_36)

  cv2.circle(frame, (x_m1, y_m1), 3,(0, 255, 0), -1)
  cv2.circle(frame, (x_m2, y_m2), 3,(0, 255, 0), -1)
 cv2.imshow("Frame", frame)
```

```
key = cv2.waitKey(1)
if key ==27:
 break
```

6.7.3 Fuzzy logic controller

The fuzzy logic controller was developed using MATLAB and the code is shown as follows:

This code can be saved as fuzzy_logic.m and run in MATLAB.

```
clear all, close all, clc

%% Generate new fuzzy
a = newfis('introbot');

%% add first input with its membership functions
a = addvar(a, 'input', 'Swab_inside_distance (m)', [0 0.07]);

a = addmf(a, 'input', 1, 'A', 'trimf', [0 0 0.005]);
a = addmf(a, 'input', 1, 'J', 'trimf', [0.0025 0.01875 0.04]);
a = addmf(a, 'input', 1, 'HW', 'trimf', [0.03 0.045 0.06]);
a = addmf(a, 'input', 1, 'WI', 'trapmf', [0.05 0.06 0.07 0.07]);

%% add second input with its membership functions
a = addvar(a, 'input', 'Angular_Position of swab (degrees)', [-90
90]);

a = addmf(a, 'input', 2, 'N', 'trapmf', [-90 -90 -20 -15]);
a = addmf(a, 'input', 2, 'SN', 'trapmf', [-20 -15 -5 -1]);
a = addmf(a, 'input', 2, 'Z', 'trimf', [-2 0 2]);
a = addmf(a, 'input', 2, 'P', 'trapmf', [1 5 15 20]);
a = addmf(a, 'input', 2, 'SP', 'trapmf', [15 20 90 90]);

%% add third input with its membership functions
a = addvar(a, 'input', 'force (N)', [0 1.1]);

a = addmf(a, 'input', 3, 'VS', 'trapmf', [0 0 0.2 0.4]);
a = addmf(a, 'input', 3, 'S', 'trapmf', [0.3 0.4 0.5 0.6]);
a = addmf(a, 'input', 3, 'SS', 'trapmf', [0.5 0.6 0.7 0.8]);
a = addmf(a, 'input', 3, 'L', 'trapmf', [0.7 0.8 0.9 1]);
a = addmf(a, 'input', 3, 'VL', 'trapmf', [0.9 1 1.1 1.1]);

%% add first output with its membership functions
a = addvar(a, 'output', 'End effector velocity (m/s)', [-0.5 0.5]);

a = addmf(a, 'output', 1, 'Zero', 'gaussmf', [0.1 0]);
a = addmf(a, 'output', 1, 'Small Positive', 'gaussmf', [0.1 0.25]);
a = addmf(a, 'output', 1, 'Positive', 'gaussmf', [0.1 0.5]);
```

```
%% add second output with its membership functions
a = addvar(a, 'output', 'Spherical wrist velocity (rad/s)', [-8
8]);
a = addmf(a, 'output', 2, 'MD', 'trapmf', [-8 -8 -6 -4]);
a = addmf(a, 'output', 2, 'D', 'trapmf', [-6 -4 -2 -1]);
a = addmf(a, 'output', 2, 'R', 'trimf', [-2 0 2]);
a = addmf(a, 'output', 2, 'U', 'trapmf', [1 2 4 6]);
a = addmf(a, 'output', 2, 'MU', 'trapmf', [4 6 8 8]);

%% Add the rules
%sys = readfis('introbot');

rule1 = [1 1 0 2 3 1 1];
rule2 = [2 1 0 3 3 1 1];
rule3 = [3 1 0 3 3 1 1];
rule4 = [4 1 0 2 3 1 1];
rule5 = [1 2 0 2 3 1 1];
rule6 = [2 2 0 3 3 1 1];
rule7 = [3 2 0 3 3 1 1];
rule8 = [4 2 0 2 3 1 1];
rule9 = [2 3 0 3 3 1 1];
rule10 = [3 3 0 2 3 1 1];
rule11 = [4 3 0 2 3 1 1];
rule12 = [3 4 0 2 3 1 1];
rule13 = [4 4 0 1 3 1 1];
rule14 = [4 5 0 1 3 1 1];
rule15 = [1 3 1 1 5 1 1];
rule16 = [1 3 2 1 3 1 1];
rule17 = [1 3 4 1 2 1 1];
rule18 = [1 3 5 1 1 1 1];
rule19 = [1 4 1 1 5 1 1];
rule20 = [1 4 2 1 3 1 1];
rule21 = [1 4 4 1 2 1 1];
rule22 = [1 4 5 1 1 1 1];
rule23 = [1 5 1 1 5 1 1];
rule24 = [1 5 2 1 3 1 1];
rule25 = [1 5 4 1 2 1 1];
rule26 = [1 5 5 1 1 1 1];

ruleList   =   [rule1;rule2;rule3;rule4;rule5;rule6;rule7;rule8;
rule9;rule10;...
 rule11;rule12;rule13;rule14;rule15;rule16;rule17;rule18;
rule19;...
```

```
 rule20;rule21;rule22;rule23;rule24;rule25;rule26];

sys = addrule(a, ruleList);

showrule(sys)

%% perform fuzzy inference calculations
end_effector_motion = evalfis([0.001 0 0.4],sys)

inputs = [0.04 0 0.88;
           0.003 -10 0.77;
  0.004 15 0.9];
end_effector_motions = evalfis(inputs,sys)

%% plot membership function and rule surface
plotmf(a, 'input',1)
figure
plotmf(a, 'input',2)
figure
plotmf(a, 'input',3)
figure
plotmf(a, 'output',1)
figure
plotmf(a, 'output',2)
figure
gensurf(sys)
```

6.8 Conclusion

Numerous difficulties in the management of COVID-19 virus for both general well-being organizations and medical clinics have been brought about by the out of the blue huge and boundless effect of the pandemic. Keeping up social distancing, lessening individual-to-individual contact, accomplishing fast diagnosis, following infection spread, and giving disinfection are a portion of the significant difficulties in the pandemic time frame. In this survey study, existing technologies adapted for battling COVID-19 were reviewed. A new application of AI and its principles during the COVID-19 pandemic was developed. It was shown that this new method used for sample collection, though a bit complex, can be very helpful in the fight against this novel virus. This is because the system is fully autonomous, therefore it will reduce the risks of healthcare workers infection, while at the same time improving the testing rate and reducing the use of PPE.

The findings from this study will contribute to the main research currently being undertaken to design an intelligent robotic machine for testing COVID-19 virus. The results will be used mainly in robotics control and design, specifically the safety aspect of the robot.

References

AAAS. (2020). *NTU Singapore researchers build disinfection robot to aid cleaners in COVID-19 outbreak*. EurekAlert! Science News. Available at: https://www.eurekalert.org/pub_releases/2020-04/ntu-nsr041420.php (Accessed 21 June 2021).

Agarwal, S., et al. (2021). *Unleashing the power of disruptive and emerging technologies amid COVID-19*. arxiv.org. Praveer Saxena. Available at: https://arxiv.org/abs/2005.11507 (Accessed 21 June 2021).

Akçakaya, M., et al. (2019). Scan-specific robust artificial-neural-networks for k-space interpolation (RAKI) reconstruction: Database-free deep learning for fast imaging. *Magnetic Resonance in Medicine*, *81*(1), 439–453. https://doi.org/10.1002/mrm.27420. John Wiley and Sons Inc.

Alkatheeri, Y., et al. (2020). The effect of big data on the quality of decision-making in Abu Dhabi government organisations. In *Advances in intelligent systems and computing* (pp. 231–248). Springer. https://doi.org/10.1007/978-981-13-9364-8_18.

Al-Sahaf, H., et al. (2019). A survey on evolutionary machine learning. *Journal of the Royal Society of New Zealand*, 205–228. https://doi.org/10.1080/03036758.2019.1609052. Taylor and Francis Asia Pacific.

Antipova, T. (2021). Coronavirus pandemic as black swan event. In *Lecture notes in networks and systems* (pp. 356–366). Springer. https://doi.org/10.1007/978-3-030-49264-9_32.

Ashima, R., Haleem, A., & Bahl, S. (2021). Automation and manufacturing of smart materials in Additive Manufacturing technologies using Internet of Things towards the adoption of Industry 4.0. *Materials Today: Proceedings*. Elsevier. Available at: https://www.sciencedirect.com/science/article/pii/S2214785321006751 (Accessed 21 June 2021).

Barfield, W. (2015). *Cyber-humans: Our future with machines*. Available at: https://books.google.com/books?hl=en&lr=&id=eVo-CwAAQBAJ&oi=fnd&pg=PR7&dq=*This+wide+acknowledgment+of+artificial+intelligence,+explicitly+in+the+field+of+medical+care+is+related+to+the+intricacies+of+the+cutting-edge+medical+services+industry+alongside+the+advancement+of+infections+which+requires+an+in-depth+data+for+diagnosis.&ots=BTUPSjV7I7&sig=1IsCck-OGYC1LAZmKsvbV-8Mkkc (Accessed 21 June 2021).

Benjamin, E. J., et al. (2018). Heart disease and stroke statistics—2018 update: A report from the American Heart Association. *Circulation*, *137*(12), E67–E492. https://doi.org/10.1161/CIR.0000000000000558. Lippincott Williams and Wilkins.

Boutaba, R., et al. (2018). A comprehensive survey on machine learning for networking: Evolution, applications and research opportunities. *Journal of Internet Services and Applications*, *9*(1). https://doi.org/10.1186/s13174-018-0087-2. Springer London.

Cavallo, A., & Rigobon, R. (2016). The billion prices project: Using online prices for measurement and research. *Journal of Economic Perspectives*, *30*(2), 151–178. https://doi.org/10.1257/jep.30.2.151. American Economic Association.

Colburn, T. (2015). *Philosophy and computer science*. Available at: https://books.google.com/books?hl=en&lr=&id=w5xzCQAAQBAJ&oi=fnd&pg=PP1&dq=*Artificial+Neural+Networks+(ANNs)+is+perhaps+the+most+noticeable+models+in+regards+to+the+role+of+artificial+intelligence+in+clinical,+as+the+greater+part+of+AI+specialists,+following+their+motivation+created+ANNs,+which+is+an+endeavor

+to+reproduce+the+ne&ots=_HcNlDxSHI&sig=oFTscnegxi0x5yMfWn_hGA9sNvc (Accessed 21 June 2021).

Constantiou, I., & Kallinikos, J. (2015). New games, new rules: Big data and the changing context of strategy. *Journal of Information Technology*, *30*(1), 44–57. https://doi.org/10.1057/jit.2014.17. Palgrave Macmillan Ltd.

Courbariaux, M., Bengio, Y., & David, J. P. (2015). Binaryconnect: Training deep neural networks with binary weights during propagations. In *Advances in neural information processing systems* (pp. 3123–3131). Neural Information Processing Systems Foundation.

Currie, G. (2019). Intelligent imaging: Anatomy of machine learning and deep learning. *Journal of Nuclear Medicine Technology*. Soc Nuclear Med. Available at: https://tech.snmjournals.org/content/47/4/273.short (Accessed 21 June 2021).

De Gruyter, W., et al. (2011). Preanalytical quality improvement: From dream to reality Chair of the EFCC Working Group on Accreditation. *Clinical Chemistry and Laboratory Medicine*, *49*(7), 1113–1126. https://doi.org/10.1515/CCLM.2011.600.

Dellinger, R. P., et al. (2008). Surviving Sepsis Campaign: International guidelines for management of severe sepsis and septic shock: 2008. *Intensive Care Medicine*, *34*(1), 17–60. https://doi.org/10.1007/s00134-007-0934-2.

Dimiduk, D. M., Holm, E. A., & Niezgoda, S. R. (2018). Perspectives on the impact of machine learning, deep learning, and artificial intelligence on materials, processes, and structures engineering. *Integrating Materials and Manufacturing Innovation*, 157–172. https://doi.org/10.1007/s40192-018-0117-8. Springer Science and Business Media Deutschland GmbH.

Dunjko, V., & Briegel, H. J. (2017). *Machine learning & artificial intelligence in the quantum domain*. Available at: http://arxiv.org/abs/1709.02779 (Accessed 21 June 2021).

Ekman, I., et al. (2011). Person-centered care—Ready for prime time. *European Journal of Cardiovascular Nursing*, *10*(4), 248–251. https://doi.org/10.1016/j.ejcnurse.2011.06.008.

Elavarasan, R. (2020). Restructured society and environment: A review on potential technological strategies to control the COVID-19 pandemic. *Science of the Total Environment*. Elsevier. Available at: https://www.sciencedirect.com/science/article/pii/S0048969720323755 (Accessed 21 June 2021).

Emakhu, J. (2019). *Failure type prediction in software-related medical device recalls*. Available at: https://search.proquest.com/openview/d680ae6f910eea8b663f019248fa11a0/1?pq-origsite=gscholar&cbl=18750&diss=y (Accessed 21 June 2021).

Flogie, A., & Aberšek, B. (2019). *The impact of innovative ICT education and AI on the pedagogical paradigm*. Available at: https://books.google.com/books?hl=en&lr=&id=yBaVDwAAQBAJ&oi=fnd&pg=PR5&dq=*By+the+progression+of+time+for+example+in+1950s,+a+reasonable+and+pragmatic+definition+with+respect+to+the+AI+innovation+was+proposed.+In+the+current+time,+the+role+and+proficiency+of+AI+has+gone+through+an+emotional+restoration+because+of+the+broad+contr&ots=hEM5ZUz4s6&sig=tusg9Df6fK_qDr_B3jx_ixZ-dFg (Accessed 21 June 2021).

Géron, A. (2019). *Hands-on machine learning with Scikit-Learn, Keras, and TensorFlow: Concepts, tools, and techniques to build intelligent systems*. Available at: https://books.google.com/books?hl=en&lr=&id=HHetDwAAQBAJ&oi=fnd&pg=PP1&dq=*Deep+learning,+a+subset+of+ML,+uses+a+progressive+degree+of+artificial+neural+networks+to+do+the+cycle+of+machine+learning.+&ots=0Lpi-thlVw&sig=ib5RymBpdMzWfzcvJ9_AB7O650I (Accessed 21 June 2021).

Haynes, R., et al. (2002). Interventions for helping patients to follow prescriptions for medications. In *Cochrane database of systematic reviews* John Wiley & Sons, Ltd. https://doi.org/10.1002/14651858.cd000011.

He, B., & Bai, K. J. (2021). Digital twin-based sustainable intelligent manufacturing: A review. *Advances in Manufacturing*, *9*(1). https://doi.org/10.1007/s40436-020-00302-5. Shanghai University.

He, J., et al. (2019). The practical implementation of artificial intelligence technologies in medicine. *Nature Medicine*. nature.com. Available at: https://www.nature.com/articles/s41591-018-0307-0 (Accessed 21 June 2021).

Henriksen, J. (2013). *Exploratory study of nursing leadership: An inquiry of leadership attributes of nurse managers in acute care hospitals and the professional practice environment*. Available at: https://search.proquest.com/openview/d2b0a2162b4e06824ecc94f119385e38/1?pq-origsite=gscholar&cbl=18750&diss=y (Accessed 21 June 2021).

Holland, J., Kingston, L., & McCarthy, C. (2021). Service robots in the healthcare sector. *Robotics*. mdpi.com. Available at: https://www.mdpi.com/1029748 (Accessed 21 June 2021).

Holland, J., et al. (2021). Service robots in the healthcare sector. *Robotics*. https://doi.org/10.3390/robotics10010047.

Keding, C. (2020). *Understanding the interplay of artificial intelligence and strategic management: Four decades of research in review*. Springer. Available at: https://link.springer.com/content/pdf/10.1007/s11301-020-00181-x.pdf. (Accessed 21 June 2021).

King, T., Cole, M., & Farber, J. (2017). *Food safety for food security: Relationship between global megatrends and developments in food safety*. Elsevier. Available at: https://www.sciencedirect.com/science/article/pii/S0924224416303703. (Accessed 21 June 2021).

Kumar, R. (2017). *Machine learning and cognition in enterprises: Business intelligence transformed*. Available at: https://books.google.com/books?hl=en&lr=&id=d5A-DwAAQBAJ&oi=fnd&pg=PR6&dq=*This+wide+acknowledgment+of+artificial+intelligence,+explicitly+in+the+field+of+medical+care+is+related+to+the+intricacies+of+the+cutting-edge+medical+services+industry+alongside+the+advancement+of+infections+which+requires+an+in-depth+data+for+diagnosis.&ots=Xbsvvhl9B6&sig=ch0xxi00saob5I51LRceyQfps3A (Accessed 21 June 2021).

Kuroi, K., & Shimozuma, K. (2004). Neurotoxicity of taxanes: Symptoms and quality of life assessment. *Breast Cancer*, 92–99. https://doi.org/10.1007/BF02968010.

Laker, L. F., et al. (2018). Quality and efficiency of the clinical decision-making process: Information overload and emphasis framing. *Production and Operations Management*, *27*(12), 2213–2225. https://doi.org/10.1111/poms.12777. Wiley-Blackwell.

Leslie, L., et al. (2020). 2020 AOA research abstracts and poster competition. *The Journal of the American Osteopathic Association*, *120*(12), e1–e87. https://doi.org/10.7556/jaoa.2020.157. NLM (Medline).

Litewka, S. G., & Heitman, E. (2020). Latin American healthcare systems in times of pandemic. *Developing World Bioethics*, *20*(2), 69–73. https://doi.org/10.1111/dewb.12262. Blackwell Publishing Ltd.

Luengo-Oroz, M., & Pham, K. (2020). Artificial intelligence cooperation to support the global response to COVID-19. *Nature*. nature.com. Available at: https://www.nature.com/articles/s42256-020-0184-3 (Accessed 21 June 2021).

Madhavan, M. V., et al. (2021). Res-CovNet: An internet of medical health things driven COVID-19 framework using transfer learning. *Neural Computing and Applications*. https://doi.org/10.1007/s00521-021-06171-8.

Magid, E., et al. (2021). Automating pandemic mitigation. *Advanced Robotics*, *35*(9), 572–589. https://doi.org/10.1080/01691864.2021.1905059. Robotics Society of Japan.

Makarius, E. E., et al. (2020). Rising with the machines: A sociotechnical framework for bringing artificial intelligence into the organization. *Journal of Business Research*. https://doi.org/10.1016/j.jbusres.2020.07.045.

Makridakis, S. (2017). *The forthcoming artificial intelligence (AI) revolution: Its impact on society and firms*. Elsevier. Available at: https://www.researchgate.net/publication/312471523. (Accessed 21 June 2021).

Mathews, S. M. (2019). Explainable artificial intelligence applications in NLP, biomedical, and malware classification: A literature review. In *Advances in intelligent systems and computing* (pp. 1269–1292). Springer Verlag. https://doi.org/10.1007/978-3-030-22868-2_90.

Matteo, M. (2012). *Development of a sustainable lean six sigma framework in healthcare sector.* Available at: http://shura.shu.ac.uk/18466/ (Accessed 21 June 2021).

McWilliams, J. M. (2020). Professionalism revealed: Rethinking quality improvement in the wake of a pandemic. *NEJM Catalyst, 1*(5). https://doi.org/10.1056/cat.20.0226. Massachusetts Medical Society.

Miranda, S. M., & Kavan, C. B. (2005). Moments of governance in IS outsourcing: Conceptualizing effects of contracts on value capture and creation. *Journal of Information Technology, 20*(3), 152–169. https://doi.org/10.1057/palgrave.jit.2000045.

Montellano, J. M. (2020). Butterfly, larvae and pupae defects detection using convolutional neural network and Apriori algorithm. In *Advances in intelligent systems and computing* (pp. 132–161). Springer. https://doi.org/10.1007/978-3-030-32523-7_10.

Mullane, & Bischofberger. (2020). *The importance of new technologies and entrepreneurship in business.* Springer.

Nammour, F. L. (2018). *Clinical intelligence framework for decision-support.* Available at: http://eprints.kingston.ac.uk/id/eprint/42973/ (Accessed 21 June 2021).

Pandey, et al. (2020). Translational genomics for achieving higher genetic gains in groundnut. *Theoretical and Applied Genetics, 133*, 1679–1702.

Press, G. (2020). *12 artificial intelligence (AI) milestones: 2. Ramon Llull and his 'thinking machine'.* Available at: https://www.forbes.com/sites/gilpress/2020/02/26/12-artificial-intelligence-ai-milestones-2-ramon-llull-and-his-thinking-machine/?sh=32825109251b (Accessed 21 June 2021).

Rojo, J. (2020). *Urban climate resilience and the promise of Big Data solutions assessing Big Data application into Madrid's urban climate change resilience scenario.* KTH Royal Institute of Technology School of Architecture and the Built Environment. Available at: https://www.diva-portal.org/smash/record.jsf?pid=diva2:1502614 (Accessed 21 June 2021).

Roth, T. L., Li, P. J., Blaeschke, F., Nies, J. F., Apathy, R., Mowery, C., … Marson, A. (2020). Pooled knockin targeting for genome engineering of cellular immunotherapies. *Cell, 181*(3), 728–744.

Shankar, K., et al. (2020). Secret image sharing scheme with encrypted shadow images using optimal homomorphic encryption technique. *Journal of Ambient Intelligence and Humanized Computing, 11*(5), 1821–1833. https://doi.org/10.1007/s12652-018-1161-0. Springer.

Sharma, P. (2021). Internet of things and blockchain. In *Blockchain for business* (pp. 296–336). Wiley. https://doi.org/10.1002/9781119711063.ch13.

Sharma, R., Mahapatra, R. P., & Agarwal, P. (2020). Transforming healthcare through various technique in Internet of Things. In *Intelligent systems reference library* (pp. 171–190). Springer Science and Business Media Deutschland GmbH. https://doi.org/10.1007/978-3-030-23983-1_7.

Shorten, C., & Khoshgoftaar, T. M. (2019). A survey on image data augmentation for deep learning. *Journal of Big Data, 6*(1). https://doi.org/10.1186/s40537-019-0197-0. SpringerOpen.

Sinclair, R. R., et al. (2021). Caught between Scylla and Charybdis: How economic stressors and occupational risk factors influence workers' occupational health reactions to COVID-19. *Applied Psychology*, 85–119. https://doi.org/10.1111/apps.12301. Blackwell Publishing Ltd.

Singh Gill, S., et al. (2019). *Transformative effects of IoT, blockchain and artificial intelligence on cloud computing: Evolution, vision, trends and open challenges.* Elsevier. Available at: https://www.sciencedirect.com/science/article/pii/S2542660519302331. (Accessed 21 June 2021).

Singh, S., Sharma, P., & Yoon, B. (2020). Convergence of blockchain and artificial intelligence in IoT network for the sustainable smart city. *Sustainable Cities and Society.* Elsevier.

Available at: https://www.sciencedirect.com/science/article/pii/S2210670720305850 (Accessed 21 June 2021).

Singla, S. (2020). Feature ranking for energy disaggregation view project Sonia Singla independent researcher AI and IoT in healthcare. In *Internet of things use cases for the healthcare industry* (pp. 1–23). Springer International Publishing. https://doi.org/10.1007/978-3-030-37526-3_1.

Skyttner, L. (2001). *General systems theory: Ideas & applications*. Available at: https://books.google.com/books?hl=en&lr=&id=TZf7bd8tguMC&oi=fnd&pg=PA3&dq=*By+the+progression+of+time+for+example+in+1950s,+a+reasonable+and+pragmatic+definition+with+respect+to+the+AI+innovation+was+proposed.+In+the+current+time,+the+role+and+proficiency+of+AI+has+gone+through+an+emotional+restoration+because+of+the+broad+contr&ots=bem743cAPa&sig=xCVr8Cupj7kYj98GiiHTpfZ9PS4 (Accessed 21 June 2021).

Solanki, S., et al. (2020). Emerging trends in microfluidics based devices. *Biotechnology Journal*. https://doi.org/10.1002/biot.201900279. Wiley-VCH Verlag.

Solima, L., Della Peruta, M. R., & Maggioni, V. (2016). Managing adaptive orientation systems for museum visitors from an IoT perspective. *Business Process Management Journal*, *22*(2), 285–304. https://doi.org/10.1108/BPMJ-08-2015-0115. Emerald Group Publishing Ltd.

Soufi, G., Hekmatnia, A., & Nasrollahzadeh, M. (2020). SARS-CoV-2 (COVID-19): New discoveries and current challenges. *Applied Sciences*. mdpi.com. Available at: https://www.mdpi.com/2076-3417/10/10/3641 (Accessed 21 June 2021).

Tande, A. J., & Patel, R. (2014). Prosthetic joint infection. *Clinical Microbiology Reviews*, *27*(2), 302–345. https://doi.org/10.1128/CMR.00111-13. American Society for Microbiology.

Thakuriah, P.(. V.)., Tilahun, N. Y., & Zellner, M. (2017). Big data and urban informatics: Innovations and challenges to urban planning and knowledge discovery. In *Springer geography* (pp. 11–45). Springer. https://doi.org/10.1007/978-3-319-40902-3_2.

Tien, J. M. (2017). Internet of things, real-time decision making, and artificial intelligence. *Annals of Data Science*, *4*(2), 149–178. https://doi.org/10.1007/s40745-017-0112-5. Springer Science and Business Media Deutschland GmbH.

Udgata, S. K., & Suryadevara, N. K. (2021). COVID-19: Challenges and advisory. In *SpringerBriefs in applied sciences and technology* (pp. 1–17). Springer. https://doi.org/10.1007/978-981-15-7654-6_1.

Van, N., et al. (2020). The role of human–machine interactive devices for post-COVID-19 innovative sustainable tourism in Ho Chi Minh City, Vietnam. *Sustainability*. mdpi.com. Available at: https://www.mdpi.com/2071-1050/12/22/9523 (Accessed 21 June 2021).

Wang, S., & Parsons, M. (2017). Augmented reality as a telemedicine platform for remote procedural training. *Sensors*. mdpi.com. Available at: https://www.mdpi.com/1424-8220/17/10/2294 (Accessed 21 June 2021).

Wiens, J., & Shenoy, E. S. (2018). Machine Learning for Healthcare: On the Verge of a Major Shift in Healthcare Epidemiology. *Clinical Infectious Diseases*, *66*, 149–153.

Wurtz, K., et al. (2019). Recording behaviour of indoor-housed farm animals automatically using machine vision technology: A systematic review. *PLoS ONE*, *14*(12). https://doi.org/10.1371/journal.pone.0226669. Public Library of Science.

Yang, X., et al. (2019). Concepts of artificial intelligence for computer-assisted drug discovery. *Chemical Reviews*, 10520–10594. https://doi.org/10.1021/acs.chemrev.8b00728. American Chemical Society.

Yannopoulos, A., Andronikou, V., & Varvarigou, T. (2008). Behavioural biometric profiling and ambient intelligence. In *Profiling the European citizen: Cross-disciplinary perspectives* (pp. 89–109). Netherlands: Springer. https://doi.org/10.1007/978-1-4020-6914-7_5.

Zekiria, Z., & Bachelor, S. (2018). *Addressing data accuracy and information integrity in mHealth using ML*. Available at: http://dro.deakin.edu.au/view/DU:30129033 (Accessed 21 June 2021).

Zeng, V., et al. (2011). De novo assembly and characterization of a maternal and developmental transcriptome for the emerging model crustacean *Parhyale hawaiensis*. *BMC Genomics, 12*. https://doi.org/10.1186/1471-2164-12-581.

Zhang, L., Wang, S., & Liu, B. (2018). Deep learning for sentiment analysis: A survey. *Wiley Interdisciplinary Reviews: Data Mining and Knowledge Discovery, 8*(4). https://doi.org/10.1002/widm.1253. Wiley-Blackwell.

Zhang, J. Y., et al. (2021). How to best protect people with diabetes from the impact of SARS-CoV-2: Report of the international COVID-19 and diabetes summit. *Journal of Diabetes Science and Technology, 15*(2), 478–514. https://doi.org/10.1177/1932296820978399. SAGE Publications Inc.

ZhenRobotics. (2020). *ZhenRobotics—Commercial floor cleaning robot, delivery robot, autonomous patrol robot*. Available at: http://www.zhenrobot.com/index_en.html (Accessed 23 June 2021).

Further reading

Kindel Media. (2021). Full shot of robot toy. In *Studies in health technology and informatics*. https://doi.org/10.3233/978-1-60750-706-2-716.

CHAPTER 7

Validation of the kinematics and dynamics models of a robotic manipulator using the MATLAB robotics toolbox

7.1 Introduction and background

When developing a prototype for any physical system, modeling can be used to study and analyze the performance of a system (Jin, 2021a). In robotics, kinematics behavior is used to model robotic systems. The main characteristic of a robotic manipulator is its mechanical structure, which is a degree of its kinematic complexity (Gong et al., 2021a). Kinematics involves the study of robot's motion without putting into consideration forces or moments that cause these motions. Therefore robot kinematics is an analytical investigation of the robotic manipulator's motion. Robotic arm's kinematic model gives a descriptive analysis of spatial movements produced by a robot, that is, association between position and orientation of robotic arm's joint variable and the end effector configuration vector (Jin, 2021b). Developing the proper kinematics models for a robotic manipulator mechanism is of paramount importance for accessing the behavior of robotic manipulators. There are generally two types of space representations employed in kinematics modeling of robotic manipulator, namely, Cartesian space and Quaternion space (Guan et al., 2021). Transformation between any two Cartesian coordinate systems comprises of rotation and translation. Rotation can thus be represented in one of the following ways: Gibbs vector, orthonormal matrices, Euler angles, Cayley-Klein parameters, Hamilton's quaternions, and Pauli spin matrices (Huang & Zou, 2021). However, the often-used representation is the homogenous transformations given by 4×4 real matrices (orthonormal matrices). Studies on kinematic modeling of serial manipulators date back to the mid-20th century. This is when a tremendous effort was done by Denavit and Hartenberg in 1955 when they developed a consistent, precise procedure of locating coordinate frames to the serial

Healthcare Systems Design of Intelligent Testing Centers
https://doi.org/10.1016/B978-0-323-99443-9.00008-5

manipulator's joints (DeBoon, 2021; Gong et al., 2021b; Perrusquia & Yu, 2021). DH notation was the name given to this approach. The subcategories of robot kinematics are forward and inverse kinematics. Forward or direct kinematics problem is straightforward as the name suggests "direct" and involves minimal complexity in equation derivations and hence manipulators always have the forward kinematics solution. Conversely, inverse kinematics problem is way more and the corresponding solution is computationally expansive and normally requires a lot of time when controlling manipulators in real time. Robotic kinematic modeling is used mainly for automating processes in industry (Boby, 2021; Perrusquía, Yu, & Li, 2021; Zhang & Peng, 2021). Due to the working environment uncertainty, industrial robots normally consist of rigid serial links fixed at the base and hence they mimic the human arm. General applications of manipulators in industries include a variety of functions, such as pick and drop, painting, gluing, and welding (Barrientos et al., 2021; Shen et al., 2021).

Robotics dynamics is study of the relationship between the accelerations produced by forces acting on robotic mechanisms (Featherstone, 2007). Dynamic modeling is essential in the design, simulation, control, and analysis of robots. In robotics, dynamic modeling involves a set of mathematical equations used to describe a robotic manipulator's dynamic behavior (Xing et al., 2021). These equations are necessary for developing a suitable control system using computer simulations after the development of physical structure of a robotic manipulator. The robotic system is usually modeled as a rigid body system. The robot dynamics is divided into two main parts, namely forward dynamics and inverse dynamics (Bamford, 2006; Ben Hamida et al., 2019; Brinker et al., 2019): forward dynamics: work out the accelerations, given the forces and inverse dynamics: provided with the accelerations, work out the forces (Righetti et al., 2011). Various methods have been developed for computing the dynamic models of robotic manipulators and these include: the Lagrange-Euler (Baressi Šegota et al., 2020; Bhattu & Kulkarni, 2021), the Recursive Lagrangian, the Generalized D'Alembert principle, and the Newton-Euler (Hwang, 2006; Khalil, 2011; Sciavicco, Siciliano, & Villani, 1995). The Lagrange-Euler and Newton-Euler are the most commonly used methods to develop the dynamic models of robotic manipulators.

7.2 Literature review

7.2.1 Inverse dynamics

The problem of inverse dynamics involves determining the robotic manipulator's joint forces and/or torques required to produce a given trajectory motion consisting of a set of accelerations, velocities, and joint positions (Kaur et al., 2021; Kim & Kim, 2021; Tsai, 2000). Inverse dynamics is mainly utilized in trajectory planning and robot control (Balafoutis & Patel, 1991). In control applications inverse dynamics is usually employed as a feedback element to calculate generalized forces needed to achieve joint accelerations by converting positions, velocities, and accelerations, calculated with regard to some desired trajectory (Tan & Potts, 1988). Trajectory planning applications use inverse dynamics models to ensure that any given trajectory can be done without the actuators' limits being exceeded (Shin & Mckay, 1986). Furthermore, inverse dynamics calculations can also be utilized as a building block for developing forward dynamics algorithms which are helpful in doing dynamic simulations of robotic manipulators.

7.2.2 Forward dynamics

Forward dynamics involves the computation of joint accelerations given the actuator forces or torques and/or any external forces or torques exerted on the end effector are known (Anne, Wilkinson, & Li, 2021). The computation of forward dynamics is mainly utilized in simulation. This is to make sure that these calculations do not satisfy the strict speed requirements of inverse dynamics applications, provided real-time simulation is not essential. Real-time simulation is often employed to develop new robot designs and control algorithms that are more economical, powerful, and flexible (Chen et al., 2021; Plancher et al., 2021).

7.2.3 Equations of motion

The general equation of motion is given as follows:

$$\tau = M(q)\ddot{q} + C(q, \dot{q}, f) \tag{7.1}$$

where q, $\dot{q}$, $\ddot{q}$, and τ are vectors of joint position, velocity, acceleration, and joint torques, respectively. Each of these variables is an n-dimensional coordinate vector, where n is the degrees of freedom of the system, that is, the number of (independent) joint variables in the system.

f denotes the end effector force, that is, an external force acting on the robot, due to interaction with the environment. Typically, f is a spatial (6D) vector and this is necessary if the robot interacts with its environment through the end effector only.

M is the joint-space inertia matrix, and it is an $n \times n$ symmetric, positive-definite matrix.

C is called the joint-space bias force.

The power delivered by τ to the system is given as $\dot{q}^T\tau$. The equation of motion is very essential in robotics calculations (Ansarieshlaghi & Eberhard, 2019; Chablat, Venkateswaran, & Boyer, 2019; Penčić et al., 2019; Sun, Angeles, & Forbes, 2019). However, there are other ways in robotics in which the equation of motion can be represented such as the following equation.

$$\tau = M(q)\ddot{q} + C(q, \dot{q})\dot{q} + G(q) + J(q)^T f \tag{7.2}$$

It can be seen from the previous equation that the bias force has been broken down into its component parts. The term $C(q, \dot{q})$ consists of the Coriolis and centrifugal forces. $G(q)$ denotes the gravitational force, and the impact of the external force is given by $J(q)^T f$. J is the Jacobian matrix of the end effector and it satisfies the following equation:

$$v = J\dot{q} \tag{7.3}$$

where v is the spatial (6D) velocity of the end effector.

7.2.4 Dynamics algorithms

7.2.4.1 Newton-Euler algorithm (RNEA)

One of the main algorithms used in robot dynamics is the recursive Newton-Euler algorithm (RNEA) (Guo et al., 2016). The recursive Newton-Euler algorithm (RNEA) algorithm solves the inverse dynamics problem. It can also be utilized to calculate C in equation of motion, and it can be adjusted so as to calculate the terms $C(q, \dot{q})$ and $G(q)$.

7.2.4.2 Lagrange-Euler formulation

Lagrange-Euler formulation is used to determine the equation of motion for robotic systems. Lagrange's equations can be stated, for an n-degree of freedom system, as follows:

$$\frac{d}{dt}\left(\frac{dT}{\partial \dot{q}_j}\right) - \frac{dT}{\partial q_j} + \frac{dV}{\partial q_j} = \tau_j^{(n)} \tag{7.4}$$

where $\dot{q}_j = \frac{\partial q}{\partial t}$ is the generalized velocity,

$\tau_j^{(n)}$ is the generalized torque on robotic joints,
q_j corresponds to generalized coordinate, and
T and V are the total kinetic and potential energy.

7.2.5 Degrees of freedom

Degrees of freedom (D.O.F), in mechanical context, are specific, characterized modes in which a mechanical device or framework can move (Rouse, 2013). The number of degrees of freedom is the same as the overall number of independent displacements or aspects of motion. The degrees of freedom of any mechanism can be calculated using Gruebler equation (Osgouie & Gard, 2017) given by the following formula:

$$D.O.F = 3(N-1) - 2J \tag{7.5}$$

where $D.O.F$ = degrees of freedom
N = number of links
J = number of joints

7.2.6 Forward kinematics

Forward kinematics determines the resultant position and orientation of the end effector after setting joints to a particular position (Owen-Hill, 2015). The commonly used approach to this problem is Denavit-Hartenberg (DH) parameters (Ferrentino & Chiacchio, 2019a, 2019b; Hill et al., 2019). Each robotic joint is characterized by four parameters. These parameters are a_{i-1}, α_{i-1}, d_i, and θ_i where the symbols represent link length, link twist, link offset, and joint angle, respectively. To determine the DH parameters a coordinate frame (see Fig. 7.1) is attached to each joint (Toussaint, 2014).

To obtain the basic transformation matrix ${}^{i-1}_{i}T$ for a single link, the following procedure is done.

$$\begin{aligned} {}^{i-1}_{i}T &= R_x(\alpha_{i-1})D_x(a_{i-1})R_z(\theta_i)Q_i(d_i) \\ &= \begin{bmatrix} \cos\theta_i & -\sin\theta_i & 0 & a_{i-1} \\ \sin\theta_i\cos\alpha_{i-1} & \cos\theta_i\cos\alpha_{i-1} & -\sin\alpha_{i-1} & -\sin\alpha_{i-1}d_i \\ \sin\theta_i\sin\alpha_{i-1} & \cos\theta_i\sin\alpha_{i-1} & \cos\alpha_{i-1} & \cos\alpha_{i-1}d_i \\ 0 & 0 & 0 & 1 \end{bmatrix} \end{aligned} \tag{7.6}$$

where R_x and R_z present rotation and D_x and Q_i denote translation. To obtain the forward kinematics of the end effector with respect to the base, multiply all of the ${}^{i-1}_{i}T$ matrices.

7.2.7 Inverse kinematics

Inverse kinematics incorporates formulating the kinematic equations essential to command a robot to a specific position within the workspace (Ferrentino & Chiacchio, 2019b; Hu et al., 2019; Otani et al., 2019; Rodríguez León et al., 2019). Inverse kinematics determines the positions robotic

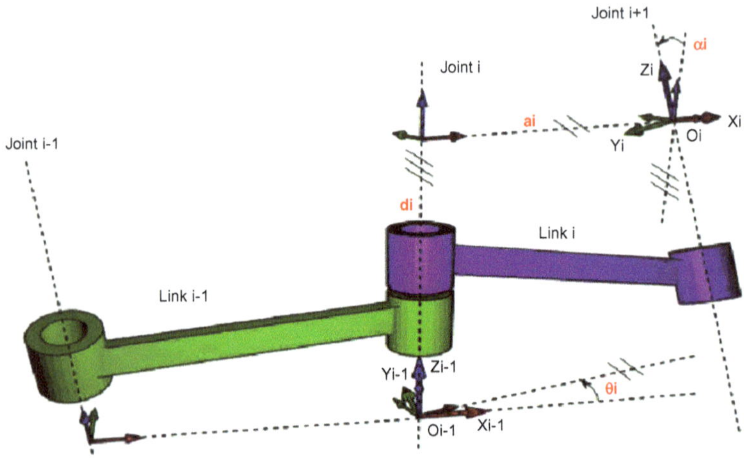

Fig. 7.1 DH parameter: Coordinate frame. *(Courtesy Hock, O., & Sedo, J. (2017).* The four parameters of classic DH convention are θ i, d i, a i, α i... *Download Scientific Diagram. Available at: https://www.researchgate.net/figure/The-four-parameters-of-classic-DH-convention-are-th-i-d-i-a-i-a-i-4_fig 8_321954826. (Accessed 14 August 2020).)*

joints should be placed at to move the end effector to a particular position and orientation (Dong et al., 2019; Fomin, Glazunov, & Terekhova, 2019; Lin & Guo, 2019; Peidró et al., 2019; Wu & Caro, 2019).

7.2.8 Robotics toolbox

MATLAB is a powerful software which allows the analysis of linear algebra and graphical presentation on a variety of computer platforms (Gil et al., 2015; Serrezuela et al., 2017). Its core functionality can be improved by installation of specific toolboxes. The robotics toolbox has many functions that are essential in robotics and solve regular problems such as dynamics, trajectory generation, and kinematics. In addition, this toolbox has graphics functions that enable the realistic representation of robotic arms. This kind of toolbox is of significance in the field of robotics where one gets a chance to practice the theory knowledge by executing practical activities (Bencomo, 2002; González, Cabrera, & Calderón, 2016; Herias Candelas & Jara, 2006). Further, in order to appreciate concepts learned theoretical, one can create their own mathematical models and simulate them. Since this robotic toolbox is an extension of MATLAB, then both simulation and modeling are done in MATLAB. Simulation in engineering field is of great significance because it enables engineers to observe real-time operation of equipment without the need of using real equipment, thereby minimizing cost and danger to human life. In addition, to simulation as well as real robots' experiment results analyzation, the robotics toolbox is an amazing tool for education.

The toolbox uses a very general approach for representing robotic manipulators' dynamics and kinematics using description matrices. Usually, this will include the robot's Denavit and Hartenberg parameters and can be developed by the operator for all serial link manipulators. Numerous examples are given for such as the well-known Puma 560 and the Stanford robotic arms (Gil et al., 2015; Righettini et al., 2015). Description of the manipulator is done through matrix augmentation, to include link, and motor inertial, as well as frictional parameters. These matrices give a precise way of robot model description and allow the sharing of robot models throughout the research community with less difficulty. Therefore this facilitates a more meaningful way of comparing simulation results than the current traditional approaches.

7.3 Materials and methods

7.3.1 Determining the DH parameters for the robotic arm

The proposed robotic manipulator consists of six links, which are all connected by revolute joints (see Fig. 7.2). A gripper has been used as the end effector and a spherical wrist connects this gripper to the manipulator.

First, we draw the axes onto each joint. The two essential axes considered are as follows:

- ✓ The z-axis which lies along the rotation axis for a revolute joint or extension axis for a prismatic joint.
- ✓ The x-axis which lies along the common normal between the previous z-axis and the current z-axis.
- ✓ The y-axis is not of importance, so the right-hand rule is used to place it.

The coloring for each axis is as follows: blue for z-axis, red for x-axis, and green for y-axis, as shown in the previous figure, see Fig. 7.3.

After that, the DH parameters are calculated in reference to the common normal.

- ✓ d is the distance measured between the current x-axis and the previous x-axis, in the previous z-axis direction.
- ✓ θ is the angle around the z-axis between the current x-axis and previous x-axis.
- ✓ a is the length of common normal.

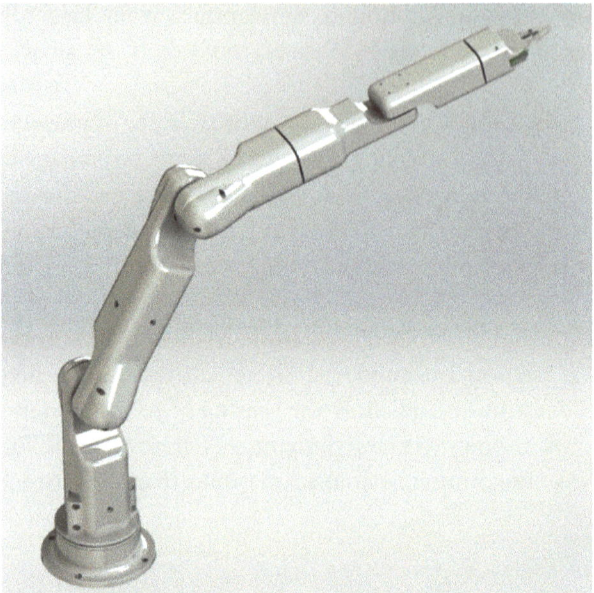

Fig. 7.2 Rendered isometric view of the proposed robotic manipulator. *(Courtesy Marvellous Moyo and Tawanda Mushiri.)*

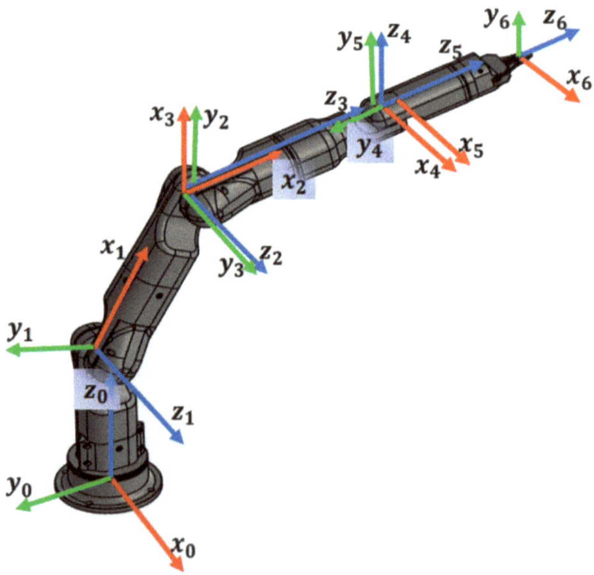

Fig. 7.3 Kinematic diagram: Frames on a robotic manipulator. *(Courtesy Marvelous Moyo and Tawanda Mushiri.)*

✓ α is the angle around the common normal between the current z-axis and previous z-axis.

Then these values are inserted on a DH parameter table for each individual link (see the Results and discussion section).

7.3.2 Determining the joint velocities

The relationship between joint velocity and end effector velocity is given as follows:

$$v = J(q)\dot{q} \tag{7.7}$$

where v is the end effector velocity,

$\dot{q}$ is the joint velocity, and

$J(q)$ is the $6 \times n$ Jacobian matrix, where n is the number of joints.

For revolute joints, the Jacobian matrix is calculated as follows:

$$[J] = \begin{bmatrix} R_{i-1}^{0}\begin{bmatrix}0\\0\\1\end{bmatrix} \times \left(d_n^0 - d_{i-1}^0\right) \\ R_{i-1}^{0}\begin{bmatrix}0\\0\\1\end{bmatrix} \end{bmatrix} \tag{7.8}$$

where R_{i-1}^{0} is the rotational matrix from the $i-1$ to 0 frame, d_n^0 is the translation matrix from frame n to frame 0, and d_{i-1}^0 is the translation from frame $i-1$ to frame 0. To obtain the joint velocities, the previous equation can be arranged such that

$$\dot{q} = [J(q)]^{-1} v \tag{7.9}$$

In this pretesticular case, the robotic manipulator has six joints and hence the order of the Jacobian matrix is 6×6. The velocity of the end effector consists of three linear and angular velocities in the x, y, z directions and therefore, it can be expressed as a 6×1 matrix. Since they are a total of six joints, the joint velocities can also be expressed as a 6×1 matrix. The column matrix for both velocities is shown as follows:

$$v = \begin{bmatrix} v_x \\ v_y \\ v_z \\ \omega_x \\ \omega_y \\ \omega_z \end{bmatrix} \text{ and } \dot{q} = \begin{bmatrix} \dot{q}_1 \\ \dot{q}_2 \\ \dot{q}_3 \\ \dot{q}_4 \\ \dot{q}_5 \\ \dot{q}_6. \end{bmatrix}$$

7.3.3 Determining the mass of each link

SolidWorks 2018 was used to model the robotic manipulator. To determine the mass of each link, a function called the mass properties, located under the evaluate section, was used.

7.3.4 Determining the robotic manipulator degrees of freedom

The proposed robotic arm has seven links, including the base link and six joints (see the earlier figure). Now, using the Gruebler equation discussed in the literature review, we have:

$$\text{D.O.F} = 3(7-1) - 2(6) = 6$$

Therefore the robotic manipulator has six degrees of freedom.

7.4 Results and discussion

7.4.1 Kinematic modeling

The kinematics of the robotic manipulator is branched into two sections: forward and inverse kinematics.

7.4.1.1 Forward kinematics

The Denavit-Hartenberg method that uses four parameters was used to compute the robotic forward kinematics. Four parameters in the table for *i*th link is a_{i-1}, α_{i-1}, d_i, and θ_i where the symbols represent link length, link twist, link offset, and joint angle, respectively (see Table 7.1).

The link transformation matrices are shown as follows:

$$ {}_{1}^{0}T = \begin{bmatrix} \cos\theta_1 & 0 & \sin\theta_1 & 0 \\ \sin\theta_1 & 0 & -\cos\theta_1 & 0 \\ 0 & 1 & 0 & a_1 \\ 0 & 0 & 0 & 1 \end{bmatrix} $$

$$ {}_{2}^{1}T = \begin{bmatrix} \cos\theta_2 & -\sin\theta_2 & 0 & a_2\cos\theta_2 \\ \sin\theta_2 & \cos\theta_2 & 0 & a_2\sin\theta_2 \\ 0 & 0 & 1 & 0 \\ 0 & 0 & 0 & 1 \end{bmatrix} $$

Table 7.1 DH parameters for the robotic manipulator.

i	a_{i-1}	α_{i-1}	d_i	θ_i
1	0	90	a_1	θ_1
2	a_2	0	0	θ_2
3	0	90	0	θ_3+90
4	0	90	a_3	θ_4+90
5	0	−90	0	θ_5
6	0	0	a_6	θ_6

$$ {}_{3}^{2}T=\begin{bmatrix} -\sin\theta_3 & 0 & \cos\theta_3 & 0 \\ \cos\theta_3 & 0 & \sin\theta_3 & 0 \\ 0 & 1 & 0 & 0 \\ 0 & 0 & 0 & 1 \end{bmatrix} $$

$$ {}_{4}^{3}T=\begin{bmatrix} -\sin\theta_4 & 0 & \cos\theta_4 & 0 \\ \cos\theta_4 & 0 & \sin\theta_4 & 0 \\ 0 & 1 & 0 & a_3 \\ 0 & 0 & 0 & 1 \end{bmatrix} $$

$$ {}_{5}^{4}T=\begin{bmatrix} \cos\theta_5 & 0 & -\sin\theta_5 & 0 \\ \sin\theta_5 & 0 & \cos\theta_5 & 0 \\ 0 & -1 & 0 & 0 \\ 0 & 0 & 0 & 1 \end{bmatrix} $$

$$ {}_{6}^{5}T=\begin{bmatrix} \cos\theta_6 & -\sin\theta_6 & 0 & 0 \\ \sin\theta_6 & \cos\theta_6 & 0 & 0 \\ 0 & 0 & 1 & a_6 \\ 0 & 0 & 0 & 1 \end{bmatrix} $$

The forward kinematics of the robotic arm can be determined by multiplying all of the ${}_{i}^{i-1}T$ matrices, where $i=1, 2, \ldots, 6$. In this case, ${}_{6}^{0}T$ is given by

$$ {}_{6}^{0}T=\begin{bmatrix} r_{11} & r_{12} & r_{13} & P_x \\ r_{21} & r_{22} & r_{23} & P_y \\ r_{31} & r_{32} & r_{33} & P_z \\ 0 & 0 & 0 & 1 \end{bmatrix} \tag{7.10} $$

Notice $\theta_{23}=\theta_2+\theta_3$

$$
\begin{aligned}
r_{11} =\ & \cos\theta_1 \sin\theta_{23}(\sin\theta_4 \cos\theta_5 \cos\theta_6 - \cos\theta_4 \sin\theta_6) \\
& + \sin\theta_1(\cos\theta_4 \cos\theta_5 \cos\theta_6 + \sin\theta_4 \sin\theta_6) \\
& + \cos\theta_1 \cos\theta_{23} \sin\theta_5 \cos\theta_6 \\
r_{12} =\ & -\cos\theta_1 \sin\theta_{23}(\sin\theta_4 \cos\theta_5 \sin\theta_6 - \cos\theta_4 \cos\theta_6) \\
& - \sin\theta_1(\cos\theta_4 \cos\theta_5 \sin\theta_6 + \sin\theta_4 \cos\theta_6) \\
& - \cos\theta_1 \cos\theta_{23} \sin\theta_5 \sin\theta_6 \\
r_{13} =\ & -\cos\theta_1 \sin\theta_{23} \sin\theta_4 \sin\theta_5 - \sin\theta_1 \cos\theta_4 \sin\theta_5 + \cos\theta_1 \cos\theta_{23} \cos\theta_5 \\
r_{21} =\ & -\sin\theta_1 \sin\theta_{23}(\sin\theta_4 \cos\theta_5 \cos\theta_6 - \cos\theta_4 \sin\theta_6) \\
& - \cos\theta_1(\cos\theta_4 \cos\theta_5 \cos\theta_6 + \sin\theta_4 \sin\theta_6) \\
& + \sin\theta_1 \cos\theta_{23} \sin\theta_5 \cos\theta_6 \\
r_{22} =\ & \sin\theta_1 \sin\theta_{23}(\sin\theta_4 \cos\theta_5 \sin\theta_6 - \cos\theta_4 \cos\theta_6) \\
& + \cos\theta_1(\cos\theta_4 \cos\theta_5 \sin\theta_6 + \sin\theta_4 \cos\theta_6) \\
& - \sin\theta_1 \cos\theta_{23} \sin\theta_5 \cos\theta_6 \\
r_{23} =\ & -\sin\theta_1 \sin\theta_{23} \sin\theta_4 \sin\theta_5 + \cos\theta_1 \cos\theta_4 \sin\theta_5 + \sin\theta_1 \cos\theta_{23} \cos\theta_5 \\
r_{31} =\ & -\cos\theta_{23}(\sin\theta_4 \cos\theta_5 \cos\theta_6 - \cos\theta_4 \sin\theta_6) + \sin\theta_{23} \sin\theta_5 \cos\theta_6 \\
r_{32} =\ & \cos\theta_{23}(\sin\theta_4 \cos\theta_5 \sin\theta_6 - \cos\theta_4 \cos\theta_6) + \sin\theta_{23} \sin\theta_5 \sin\theta_6 \\
r_{33} =\ & \cos\theta_{23} \sin\theta_4 \sin\theta_5 + \sin\theta_{23} \cos\theta_5
\end{aligned} \tag{7.11}
$$

Assuming that the spherical wrist is responsible for only orienting the end effector, then

$$a_6 = 0$$

$$P_x = a_3 \cos\theta_1 \cos\theta_{23} + a_2 \cos\theta_1 \cos\theta_2$$

$$P_y = a_3 \sin\theta_1 \cos\theta_{23} + a_2 \sin\theta_1 \cos\theta_2$$

$$P_z = a_3 \sin\theta_{23} + a_2 \sin\theta_2 + \mathrm{a}_1$$

7.4.1.2 Inverse kinematics

Denotes the coordinates of the wrist points as (x, y, z), therefore from Eq. (7.8)

$$x = a_3 \cos\theta_1 \cos\theta_{23} + a_2 \cos\theta_1 \cos\theta_2$$
$$y = a_3 \sin\theta_1 \cos\theta_{23} + a_2 \sin\theta_1 \cos\theta_2 \quad (7.12)$$
$$z = a_3 \sin\theta_{23} + a_2 \sin\theta_2 + a_1$$

using the wrist coordinates and ${}^{0}_{6}T$ (see Eq. 7.8), the joint angles can be found and they are all listed as follows:

$$\theta_2 = Atan2(s_2, c_2) \mp Atan2\left(\sqrt{s_2^2 + c_2^2 - m_2^2}, m_2\right) \quad (7.13)$$

where $s_2 = z - a_1$

$c_2 = \sqrt{x^2 + y^2}$

$m_2 = \dfrac{s_2{}^2 + c_2{}^2 + a_2^2 - a_3^2}{2a_2}$

$$\theta_{23} = Atan2\left(s_{23}, \mp\sqrt{1 - s_{23}^2}\right) \quad (7.14)$$

where $s_{23} = \dfrac{s_2 - a_2 \sin\theta_2}{a_3}$

$$\theta_3 = \theta_{23} - \theta_2 \quad (7.15)$$

$$\theta_1 = Atan2\left(s_1, \mp\sqrt{1 - s_1^2}\right) \quad (7.16)$$

where $s_1 = \dfrac{y}{a_3 \cos\theta_{23} + a_2 \cos\theta_2}$

$$\theta_4 = Atan2(-c_4, s_4) \text{ or } \theta_4 = Atan2(c_4, -s_4) \quad (7.17)$$

where $s_4 = r_{23}\cos^2\theta_1 - r_{13}\sin\theta_1\cos\theta_1$

$c_4 = r_{23}\sin\theta_1\cos\theta_1\sin\theta_{23} - r_{33}\cos\theta_1\cos\theta_{23} + r_{13}\sin\theta_{23}\cos^2\theta_1$

$$\theta_5 = Atan2\left(s_5, \mp\sqrt{1 - s_5^2}\right) \quad (7.18)$$

where $s_5 = \dfrac{r_{23}\cos\theta_1 - r_{13}\sin\theta_1}{\cos\theta_4}$

$$\theta_6 = Atan2(s_6, c_6) \mp Atan2\left(\sqrt{s_6^2 + c_6^2 - m_6^2}, m_6\right) \quad (7.19)$$

where $s_6 = \cos\theta_{23}\cos\theta_4\sin\theta_6$

$c_6 = \sin\theta_{23}\sin\theta_5 - \cos\theta_{23}\sin\theta_4\cos\theta_5$

$m_6 = r_{31}$

7.4.1.3 Robotic manipulator's workspace

The workspace of the robotic arm is the set of values for which the last link can be positioned and oriented such that the inverse kinematics solution exists. The bounding surface(s) of the 3D solid which the arm can reach is obtained by squaring and adding the three equations denoting the wrist configuration and for simplicity, assuming that the wrist is intersecting, that is, $a_6=0$. Therefore, simplifying we have:

$$R^2 = x^2 + y^2 + z^2 = K_1 + K_2 \cos\theta_3 + K_3 \sin\theta_3 \tag{7.20}$$

where $K_1 = a_1^2 + a_2^2 + a_3^2 + 2a_1a_2 \sin\theta_2$

$$K_2 = 2a_3a_2 + 2a_3a_1 \sin\theta_2$$

$$K_3 = 2a_3a_1 \cos\theta_2$$

Any point within the workspace envelope satisfies, $\frac{\partial R^2}{\partial \theta_3} = 0$ (7.21)

$$\therefore -K_2 \sin\theta_3 + K_3 \cos\theta_3 = 0 \tag{7.22}$$

Eliminating θ_3 from these two equations before, we have:

$$\left(x^2 + y^2 + z^2 - R_1\right)\left(x^2 + y^2 + z^2 + R_1\right) = 0 \tag{7.23}$$

where $R_1 = \frac{K_1\sqrt{K_2^2 + K_3^2} + (K_2^2 + K_3^2)}{\sqrt{K_2^2 + K_3^2}}$

Therefore, from the previous equation, it shows that the bounding surfaces for the wrist points are spheres.

7.4.1.4 Numerical validation using MATLAB robotic toolbox

The numerical values for the DH parameters and the range of joint angles are presented in Table 7.2.

Table 7.2 Numerical DH parameter table.

i	a_{i-1} meters	α_{i-1} degrees	d_i meters	Range of θ_i degrees
1	0	90	0.287	$-180 \leq \theta_1 \leq 180$
2	0.328	0	0	$-15 \leq \theta_2 \leq 180$
3	0	90	0	$0 \leq \theta_3 \leq 180$
4	0	90	0.318	$-180 \leq \theta_4 \leq 180$
5	0	−90	0	$-90 \leq \theta_5 \leq 90$
6	0	0	0.53	$-180 \leq \theta_6 \leq 180$

Consider an arbitrary chosen nominal position of the arm, that is, the arm dexterous pose, where $\theta_1 = 0, \theta_2 = \frac{\pi}{20}, \theta_3 = \frac{2\pi}{3}, \theta_4 = 0, \theta_5 = \frac{\pi}{6}, \theta_6 = 0$ units are in radians.

The transformation matrix ${}^0_6T$ is given as follows:

$$
{}^0_6T = \begin{bmatrix} -0.1564 & 0.0001 & 0.9877 & 0.821 \\ 0 & -1 & 0.0001 & 0 \\ 0.9877 & 0 & 1.564 & 0.578 \\ 0 & 0 & 0 & 1 \end{bmatrix}
$$

Using the earlier transformation matrix as an input to the inverse kinematics algorithm and by solving the kinematic equations, one set of the many available solutions in radians is as follows:

$$
\theta_1 = 0,\ \theta_2 = 0.6724,\ \theta_3 = 1.0472,\ \theta_4 = 0.0066,\ \theta_5 = -0.0083,\ \theta_6 = -0.0068
$$

And also, the initial position stated before will be one of the many available solutions (see the programming code at the chapter end).

Manipulator velocity

The Jacobian matrix with respect to the world coordinate frame is given as follows:

$$
[J] = \begin{bmatrix} 0 & 0.291 & -2.2397 & 0 & -0.0396 & 0 \\ 0.821 & -0.0002 & -0.0001 & 0.1265 & 0 & 0 \\ 0 & 0.821 & 0.497 & 0 & -0.2499 & 0 \\ 0 & 0 & 0 & 0.7771 & -0.0003 & 0.9877 \\ 0 & -1 & -1 & 0.0001 & 1 & 0.0001 \\ 1 & -0.0002 & -0.0002 & 0.6293 & -0.0001 & 0.1564 \end{bmatrix}
$$

Assuming the maximum allowable end effector linear velocity is 500 mm/s, therefore from the previous equation the joint velocities are obtained as follows:

$$
\begin{bmatrix} \dot{q}_1 \\ \dot{q}_2 \\ \dot{q}_3 \\ \dot{q}_4 \\ \dot{q}_5 \\ \dot{q}_6 \end{bmatrix} = \begin{bmatrix} 0 & 1.7551 & 0.0004 & 0.0693 & 0.0001 & -0.4376 \\ 4.7387 & 0.0001 & 3.8373 & -0.0001 & 0.7713 & 0.0001 \\ -10.9506 & -0.0003 & -4.8212 & 0.0001 & -0.7713 & -0.0003 \\ -0.0025 & -3.4589 & -0.0011 & 0.4495 & -0.0002 & 2.8397 \\ -6.2119 & -0.0002 & -0.9839 & -0.0001 & 1 & -0.0001 \\ 0 & 2.7216 & 0.0006 & 1.3664 & 0.0005 & -2.2344 \end{bmatrix} \begin{bmatrix} 0.5 \\ 0.5 \\ 0.5 \\ 0.3 \\ 0.3 \\ 0.3 \end{bmatrix}
= \begin{bmatrix} 0.7652 \\ 4.5195 \\ -8.1174 \\ -1.0144 \\ -3.2995 \\ 1.1008 \end{bmatrix}
$$

Therefore the joint velocities of links 1, 2, 3, 4, 5, and 6 are 0.77 rad/s, 4.52 rad/s, −8.11 rad/s, −1.01 rad/s, −3.30 rad/s, and 1.10 rad/s, respectively. This shows that to achieve this end effector motion, the second and third joint will have to rotate at very high speeds. High speeds of robotic joints occur at singularity that is when the determinant of the Jacobian matrix is zero. In this particular case, the determinant is 0.0149 and hence these joint velocities are acceptable for this configuration. Figs. 7.4 and 7.5 show the configuration of the arm at zero angle position and nominal angle position.

7.4.2 Dynamic modeling

Firstly, the mass of each link is determined using the density of the selected material for corresponding links. The material selected for each link was Kevlar 53%. Table 7.3 presents the respective mass of each link.

A total of 2 kg has been added to the end effector mass which is the maximum payload of the end effector. In addition, a scale factor of 2 was used to calculate the total volume of link 5, since the gripper mechanism inside that link was excluded during volumetric calculation. Table 7.4 presents the model parameters and their corresponding value for each link, where

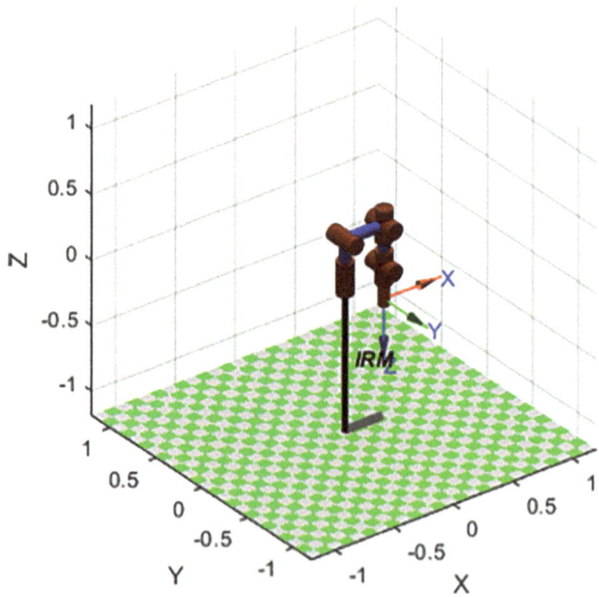

Fig. 7.4 Robotic arm configuration at zero angle position. *(Courtesy Marvellous Moyo and Tawanda Mushiri.)*

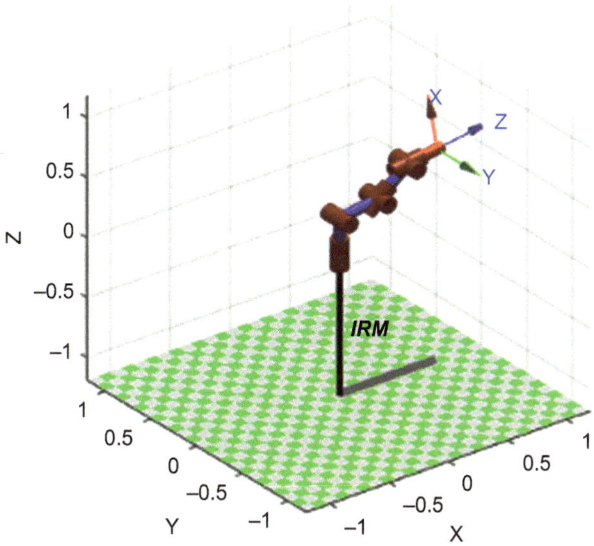

Fig. 7.5 Robotic arm configuration at nominal position. *(Courtesy Marvellous Moyo and Tawanda Mushiri.)*

Table 7.3 Volumetric properties of links.

Part	Volume (m^3)	Density (kg/m^3)	Mass (kg)
Link 6 (Base link)	0.000967	1350	1.31
Link 1	0.000910	1350	1.23
Link 2	0.000452	1350	0.61
Link 3	0.000343	1350	0.46
Link 4	0.000255	1350	0.34
Link 5 (End effector)	0.000176	1350	2.24

Table 7.4 Dynamic model parameters.

Part	*R* (m)	*H* (m)	*LX* (m)	*LY* (m)	*LZ* (m)	*b*
Link 6 (Base link)	0.06	0.287	–	–	–	1
Link 1	–	–	0.08	0.328	0.08	1
Link 2	–	–	0.08	0.149	0.08	1
Link 3	–	–	0.08	0.169	0.08	1
Link 4	–	–	0.06	0.129	0.06	1
Link 5 (End effector)	–	–	0.06	0.124	0.06	1

✓ R is the radius of the cylindrical link
✓ H is the height of the cylindrical link
✓ LX is the length of link in the x-direction
✓ LY is the length of link in the y-direction
✓ LZ is the length of link in the z-direction
✓ b is the damping coefficient

The Euler-Lagrange method has been used to derive all of the equations of motion of the robotic manipulator system. Following is a free body diagram (see Fig. 7.6) of the robotic manipulator showing how all the links are connected and types of forces acting in each joint.

Let τ_m = Actuation torque produced by each motor and

$b\dot{\theta}$ = Viscous damping torque

The general form of the equation of motion will be given as follows and the symbols have the usual meanings:

$$M(q,\dot{q})\ddot{q} + C(q,\dot{q})\dot{q} + G(q) = Q(\tau,\dot{q})$$

The length of each link is defined with respect to the x, y, z plane. From the following free body diagram, Fig. 7.7, it can be shown that the longest length will be along the y plane.

The robotic manipulator consists of five rectangular prisms which form the links of the arm. In addition, the robotic arm has a sixth link which is a

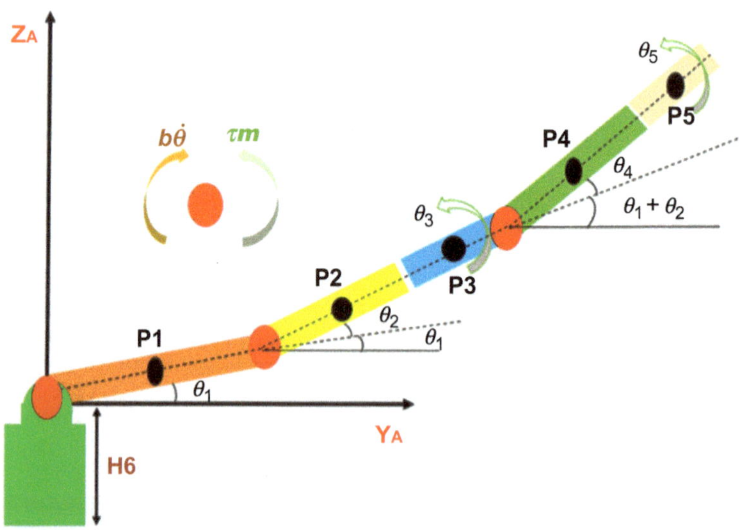

Fig. 7.6 Free body diagram of the robotic manipulator. *(Courtesy Marvellous Moyo and Tawanda Mushiri.)*

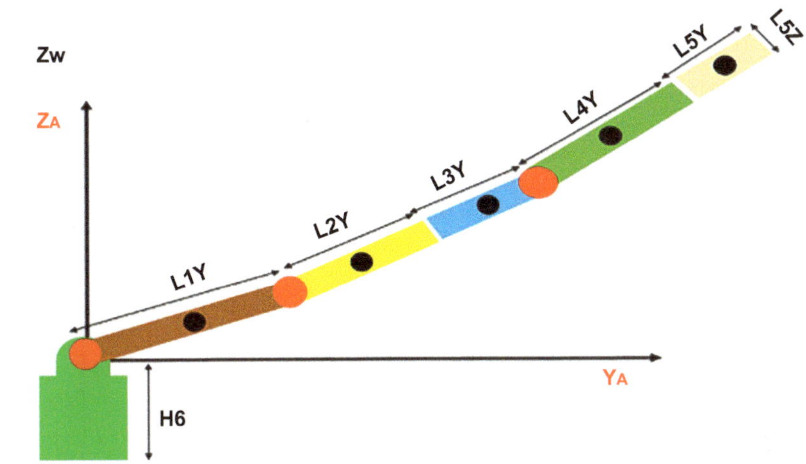

Fig. 7.7 Free body diagram of the arm showing length sizes. *(Courtesy Marvellous Moyo and Tawanda Mushiri.)*

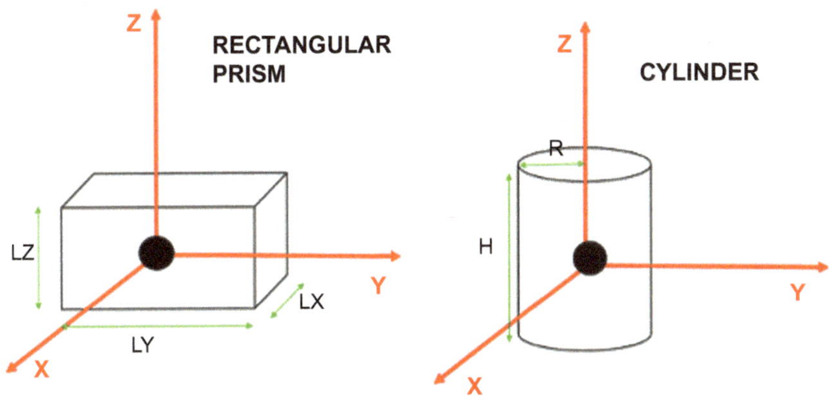

Fig. 7.8 Body frame of rectangular prism and cylinder position at the center of mass. *(Courtesy Marvellous Moyo and Tawanda Mushiri.)*

cylindrical base. Each link is considered to have a body frame attached at its center of mass and the inertia will be calculated with respect to that center (see Fig. 7.8).

Here is an example of the inertia of link 1 and the base link computed using MATLAB symbolic toolbox.

$$I_1 = \begin{pmatrix} \frac{m_{1,s}\left(\mathrm{L1Y}_s^2 + \mathrm{L1Z}_s^2\right)}{12} & 0 & 0 \\ 0 & \frac{m_{1,s}\left(\mathrm{L1X}_s^2 + \mathrm{L1Z}_s^2\right)}{12} & 0 \\ 0 & 0 & \frac{m_{1,s}\left(\mathrm{L1X}_s^2 + \mathrm{L1Y}_s^2\right)}{12} \end{pmatrix}$$

$$I_6 = \begin{pmatrix} \frac{m_{6,s}\left({H_{6,s}}^2 + 3{R_{6,s}}^2\right)}{12} & 0 & 0 \\ 0 & \frac{m_{6,s}\left({H_{6,s}}^2 + 3{R_{6,s}}^2\right)}{12} & 0 \\ 0 & 0 & \frac{{R_{6,s}}^2 m_{6,s}}{2} \end{pmatrix}$$

Next, the center of mass positions and translational velocities for the arm links will be defined. Let the inertial world coordinate (W) frame be at the pivot of the base link and link 1. The W-frame is fixed in space so it does not have any motion. Consider also frame A that has the same origin as the W-frame; however, the A-frame can rotate about the Z_W and Z_A axes (see Fig. 7.9).

From the previous diagrams, the position of each link about the A-frame at the center of gravity can be computed as follows:

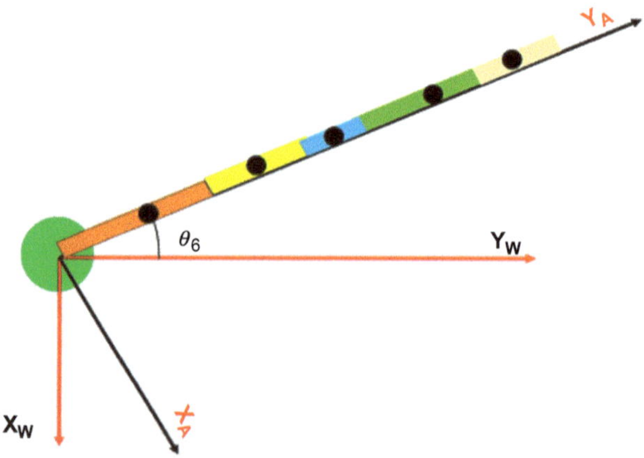

Fig. 7.9 The W-Frame and A-Frame viewed from the top. *(Courtesy Marvellous Moyo and Tawanda Mushiri.)*

$$P_1 = \begin{bmatrix} 0 \\ \frac{LY_1}{2}\cos\theta_1 \\ \frac{LY_1}{2}\sin\theta_1 \end{bmatrix}$$

Let $\alpha = \theta_1 + \theta_2$, see earlier figures

$$P_2 = \begin{bmatrix} 0 \\ LY_1\cos\theta_1 + \frac{LY_2}{2}\cos\alpha \\ LY_1\sin\theta_1 + \frac{LY_2}{2}\sin\alpha \end{bmatrix}$$

$$P_3 = \begin{bmatrix} 0 \\ LY_1\cos\theta_1 + \left(LY_2 + \frac{LY_3}{2}\right)\cos\alpha \\ LY_1\sin\theta_1 + \left(LY_2 + \frac{LY_3}{2}\right)\sin\alpha \end{bmatrix}$$

Let $\beta = \theta_1 + \theta_2 + \theta_3$

$$P_4 = \begin{bmatrix} 0 \\ LY_1\cos\theta_1 + (LY_2 + LY_3)\cos\alpha + \frac{LY_4}{2}\cos\beta \\ LY_1\sin\theta_1 + (LY_2 + LY_3)\sin\alpha + \frac{LY_4}{2}\sin\beta \end{bmatrix}$$

$$P_5 = \begin{bmatrix} 0 \\ LY_1\cos\theta_1 + (LY_2 + LY_3)\cos\alpha + \left(LY_4 + \frac{LY_5}{2}\right)\cos\beta \\ LY_1\sin\theta_1 + (LY_2 + LY_3)\sin\alpha + \left(LY_4 + \frac{LY_5}{2}\right)\sin\beta \end{bmatrix}$$

Here is coordinate P_5 obtained using MATLAB, note LY_4 corresponds to the L4Y and so does the other lengths.

$$P_5 = \begin{pmatrix} \cos(\sigma_1)\left(\mathrm{L4Y}_s + \frac{\mathrm{L5Y}_s}{2}\right) + \cos(\theta_1(t) + \theta_2(t))(\mathrm{L2Y}_s + \mathrm{L3Y}_s) + \mathrm{L1Y}_s\cos(\theta_1(t)) \\ \mathrm{L1Y}_s\sin(\theta_1(t)) + \sin(\sigma_1)\left(\mathrm{L4Y}_s + \frac{\mathrm{L5Y}_s}{2}\right) + \sin(\theta_1(t) + \theta_2(t))(\mathrm{L2Y}_s + \mathrm{L3Y}_s) \end{pmatrix}$$

where

$$\sigma_1 = \theta_1(t) + \theta_2(t) + \theta_4(t)$$

Next, convert the A-frame coordinates to corresponding W-frame coordinates using the following transformation matrix (see the top view of the model from the earlier figures).

$$ {}_{A}^{W}R = \begin{pmatrix} \cos(\theta_6(t)) & -\sin(\theta_6(t)) & 0 \\ \sin(\theta_6(t)) & \cos(\theta_6(t)) & 0 \\ 0 & 0 & 1 \end{pmatrix} $$

The new coordinates will be given by the following formulae, for link 5:

$$P_5w = {}_{A}^{W}R \cdot P_5 =$$

$$\begin{pmatrix} -\sin(\theta_6(t))\sigma_1 \\ \cos(\theta_6(t))\sigma_1 \\ \mathrm{L1Y}_s\sin(\theta_1(t)) + \sin(\theta_1(t)+\theta_2(t)+\theta_4(t))\left(\mathrm{L4Y}_s + \dfrac{\mathrm{L5Y}_s}{2}\right) + \\ \sin(\theta_1(t)+\theta_2(t))(\mathrm{L2Y}_s + \mathrm{L3Y}_s) \end{pmatrix}$$

where

$$\sigma_1 = \cos(\theta_1(t) + \theta_2(t) + \theta_4(t))\left(\mathrm{L4Y}_s + \frac{\mathrm{L5Y}_s}{2}\right) + \cos(\theta_t(t) + \theta_2(t))(\mathrm{L2Y}_s + \mathrm{L3Y}_s) + \mathrm{L1Y}_s\cos(\theta_1(t))$$

The translation velocity is obtained by using the following formulae:

$$v_5 = \frac{d}{dt}P_5w =$$

$$\begin{pmatrix} \sin(\theta_6(t))\sigma_1 - \cos(\theta_6(t))\sigma_2\dfrac{\partial}{\partial t}\theta_6(t) \\ -\cos(\theta_6(t))\sigma_1 - \sin(\theta_6(t))\sigma_2\dfrac{\partial}{\partial t}\theta_6(t) \\ \mathrm{L1Y}_s\cos(\theta_1(t))\dfrac{\partial}{\partial t}\theta_1(t) + \sigma_6\sigma_4(\mathrm{L2Y}_s + \mathrm{L3Y}_s) + \cos(\sigma_5)\left(\mathrm{L4Y}_s + \dfrac{\mathrm{L5Y}_s}{2}\right)\sigma_3 \end{pmatrix}$$

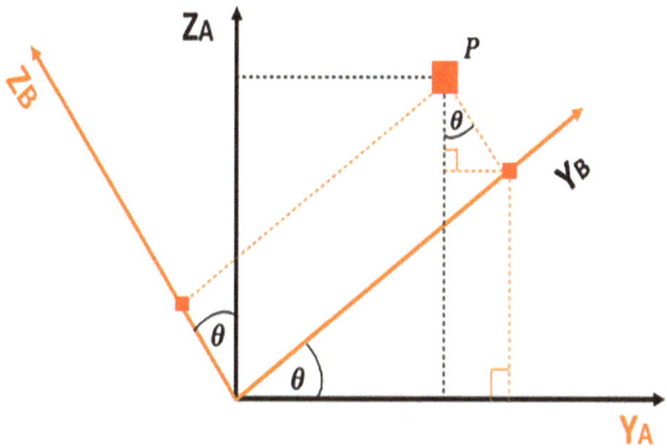

Fig. 7.10 Schematic diagram showing transformation of point P from A- to B-frame. *(Courtesy Marvellous Moyo and Tawanda Mushiri.)*

where

$$\sigma_1 = \mathrm{L1Y}_s \sin(\theta_1(t)) \frac{\partial}{\partial t}\theta_1(t) + \sin(\theta_1(t)+\theta_2(t))\sigma_4(\mathrm{L2Y}_s + \mathrm{L3Y}_s) + \sin(\sigma_5)\left(\mathrm{L4Y}_s + \frac{\mathrm{L5Y}_s}{2}\right)\sigma_3$$

$$\sigma_2 = \cos(\sigma_5)\left(\mathrm{L4Y}_s + \frac{\mathrm{L5Y}_s}{2}\right) + \sigma_6(\mathrm{L2Y}_s + \mathrm{L3Y}_s) + \mathrm{L1Y}_s \cos(\theta_1(t))$$

$$\sigma_3 = \frac{\partial}{\partial t}\theta_1(t) + \frac{\partial}{\partial t}\theta_2(t) + \frac{\partial}{\partial t}\theta_4(t)$$

$$\sigma_4 = \frac{\partial}{\partial t}\theta_1(t) + \frac{\partial}{\partial t}\theta_2(t)$$

$$\sigma_5 = \theta_1(t) + \theta_2(t) + \theta_4(t)$$

$$\sigma_6 = \cos(\theta_1(t) + \theta_2(t))$$

The angular velocities of each link can be obtained in reference to the B-frame. A transformation matrix is used to transform coordinates from A- to B-frame (see Fig. 7.10).

This transformation matrix depends on the angular rotation of each link with respect to the horizontal, which is given as follows:

$$^{B}_{A}R(\theta) = \begin{bmatrix} 1 & 0 & 0 \\ 0 & \cos\theta & \sin\theta \\ 0 & -\sin\theta & \cos\theta \end{bmatrix}$$

The rotational velocities of all the links depend on the velocity of base link given by the formulae, for example for link 5 the velocity is given as follows:

$$\omega_5 = {}^{B}_{A}R(\theta_{124}) \cdot \frac{d}{dt}\theta_6 + \frac{d}{dt}\theta_{124}$$

where θ_6 and θ_{124} are 3×1 matrix and

$\theta_{124} = \theta_1 + \theta_2 + \theta_4$

$$\omega_5 = \begin{pmatrix} \frac{\partial}{\partial t}\theta_1(t) + \frac{\partial}{\partial t}\theta_2(t) + \frac{\partial}{\partial t}\theta_4(t) \\ \sin(\theta_1(t) + \theta_2(t) + \theta_4(t))\frac{\partial}{\partial t}\theta_6(t) + \frac{\partial}{\partial t}\theta_5(t) \\ \cos(\theta_1(t) + \theta_2(t) + \theta_4(t))\frac{\partial}{\partial t}\theta_6(t) \end{pmatrix}$$

The total kinetic energy for link 5 can now be computed as follows:

$$KE_5 = \frac{1}{2}v_5{}^T \cdot m_5 \cdot v_5 + \frac{1}{2}\omega_5{}^T \cdot I_5 \cdot \omega_5$$

The corresponding potential energy will be given as follows:

$$PE_5 = m_5 g \cdot P_5 w$$

Link 5 has been used as an example for almost all the calculations but the same can be done for all the other links and the total kinetic and potential energies obtained in similar manner. Thereafter, the Lagrangian can be computed as follows:

$L = KE - PE$, where KE and PE are the total kinetic and potential energy, respectively.

The generalized force Q_k acting on the system is given as follows:

$$Q_k = \sum_{j=1}^{nf}\left(\vec{F} \cdot \frac{\partial \vec{v}}{\partial \dot{q}_k}\right) + \sum_{j=1}^{n\tau}\left(\vec{\tau} \cdot \frac{\partial \vec{\omega}}{\partial \dot{q}_k}\right)$$

where $\vec{v}$ is the velocity vector of the point associated with the applied force $\vec{F}$

$\vec{\omega}$ is the angular velocity vector of the point associated with the applied torque $\vec{\tau}$

nf is the number of active nonconservative forces, in this case equals 6

$n\tau$ is the number of active nonconservative torques, also equals six for this model.

Q_k is the generalized force associated with the kth generalized coordinate q_k

For example, the generalized force for link 5 is given as $\tau_m - b\dot{\theta}_5$. The equation of motion can then be formulated using the following formula (see Chapter 2 for better insight):

$$\frac{d}{dt}\left(\frac{\partial L}{\partial \dot{q}_k}\right) - \frac{\partial L}{\partial q_k} = Q_k$$

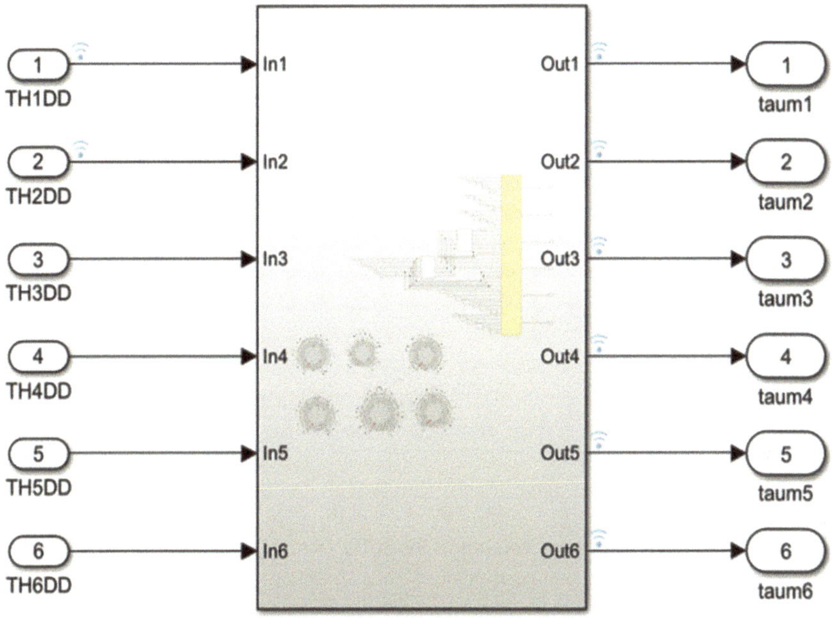

Fig. 7.11 Simulink model of the robotic manipulator dynamics. *(Courtesy Marvellous Moyo and Tawanda Mushiri.)*

Using MATLAB this equation of motion has been converted into a Simulink model block and inverse dynamics will be done for the system to determine the generalized force acting at each joint. The angular accelerations at each joint are given as TH1DD, TH2DD, TH3DD, TH4DD, TH5DD, and TH6DD, respectively (see Fig. 7.11).

7.4.2.1 Dynamic numerical solution

The earlier Simulink model has been excited with a sine wave input with a frequency of 1 rad/s. The amplitude of each input was obtained from the velocity analysis results. This was the maximum joint velocity required to reach 500 mm/s end effector linear motion. The output obtained needed some varying of input variables such as damping torque, but the maximum amplitude was recorded for corresponding inputs. Following are the graphs showing the input to the system, output before varying the input variables, output after varying input variables, and the comparison between the output and input variables, respectively (see Figs. 7.12–7.15). Note the horizontal axis represents time and vertical represents the torque amplitude in Newton meters.

From the results of the simulation, the dynamics torque for each joint was obtained as presented in Table 7.5.

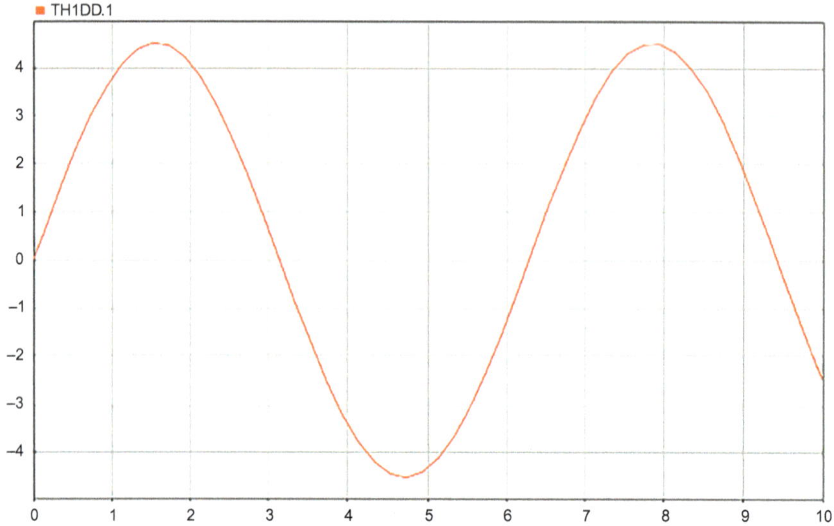

Fig. 7.12 Sine wave angular acceleration input for joint 1. *(Courtesy Marvellous Moyo and Tawanda Mushiri.)*

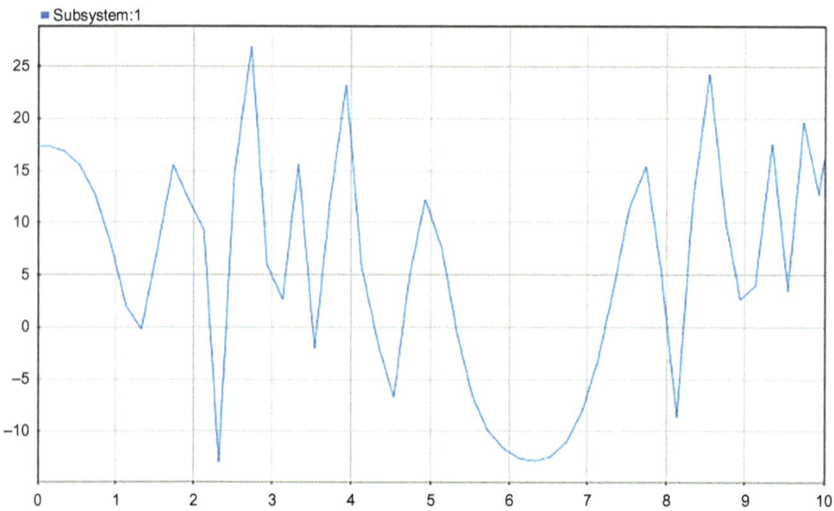

Fig. 7.13 Output torque for joint 1. *(Courtesy Marvellous Moyo and Tawanda Mushiri.)*

7.4.2.2 Motor sizing and selection

The static torques can be calculated on the assumption that the robotics links are stretched horizontal and the robot is stationary. These are the torques required to hold the robotic links against gravity. For a robotic link with length L (see Fig. 7.16 for a link with a revolute joint), the torque is calculated as follows:

$$T = I\alpha + mg\frac{L}{2} + m_e gL$$

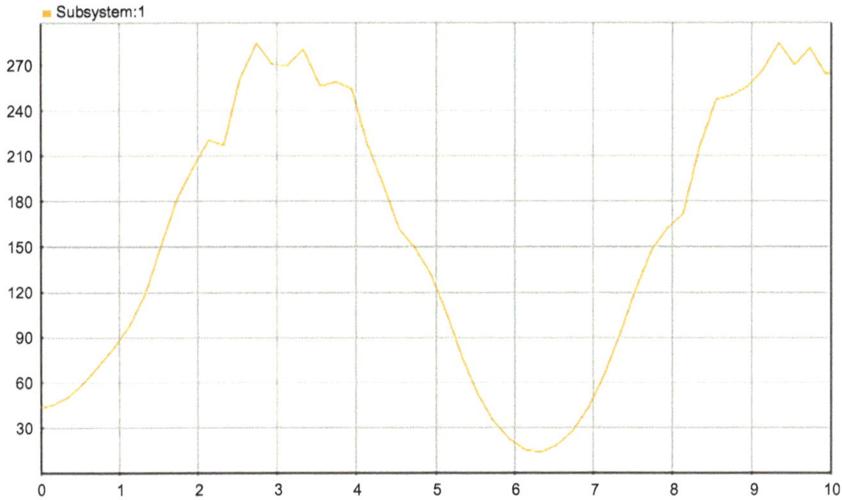

Fig. 7.14 Output torque for joint 1 after varying the input variables. *(Courtesy Marvellous Moyo and Tawanda Mushiri.)*

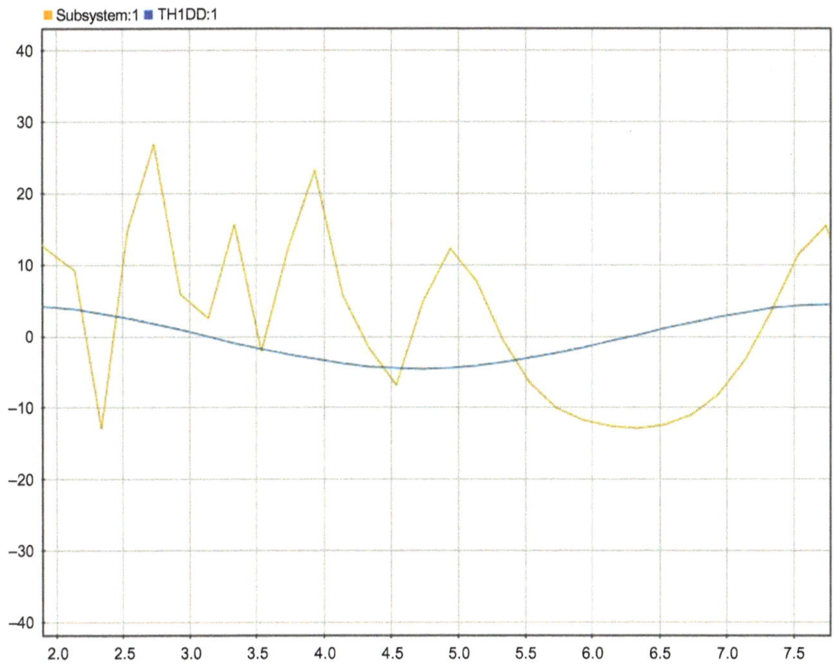

Fig. 7.15 Comparison of the input and output for joint 1, blue is the input. *(Courtesy Marvellous Moyo and Tawanda Mushiri.)*

Table 7.5 Joint dynamics torques.

Part	Torque (N m)
Joint 1	29
Joint 2	31.5
Joint 3	3.1
Joint 4	11.4
Joint 5	5
Joint 6 (base link)	6.1

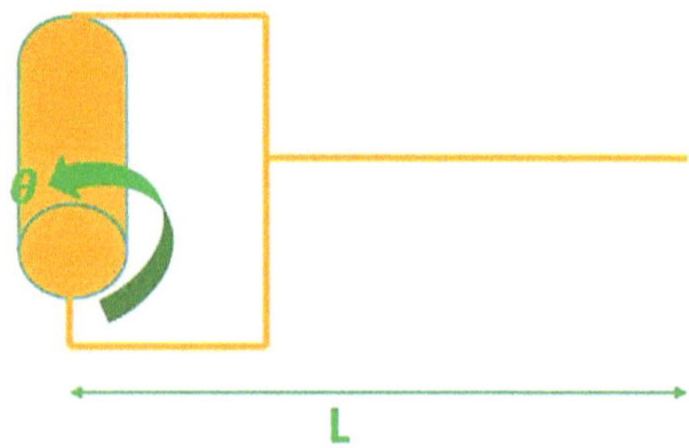

Fig. 7.16 Free body diagram of a revolute joint. *(Courtesy Marvellous Moyo and Tawanda Mushiri.)*

where m_e is the mass of end effector or preceding links.

T = torque

I = Moment of inertia $= \frac{mL^2}{3}$

α = Angular acceleration

The maximum angular velocity for the joint required to reach the maximum end effector linear velocity of 500 mm/s was −8.11 rad/s. There the angular acceleration will be approximately 6π rad/s^2 and each joint is assumed to have this acceleration during calculation. For the previous given dimensions, the torques obtained for each joint are given as presented in Table 7.6.

Table 7.6 Joint static torques.

Part	Torque (N m)
Joint 1	14.55
Joint 2	4.97
Joint 3	2.04
Joint 4	3.53
Joint 5	0.051
Joint 6 (base link)	1.81

The total torque required to actuate joints equals the sum of the dynamic and static torque and this will be compared against motor specifications in the market. For simplicity, servo motors will be used for all of the joints. In addition, to compensate for the actuators and other electrical components within robotic links, a safety factor of 2 will be used. Note should be taken that the motors selected for joint 3 and the base joint have been oversized even more compared to their respective joint torques. This is to cater for other joint poses besides the nominal pose which might be closer to singularity, leading to high joint angular velocities and in addition, the base has to rotate the whole robotic assembly. This has been done for joint 3 also, because it is like a base for the spherical wrist which is responsible for orientation of the end effector. Table 7.7 presents the sizes of motors selected for each individual links and the motor specifications are presented in Appendix B.

7.5 Programming examples

The MATLAB code used to develop the forward kinematic model of the robotic manipulator is shown as follows:

This code can be saved as forward_kinematics.m and run in MATLAB.

Table 7.7 Motor selection.

Part	Maximum torque (N m)	Maximum velocity (rad/s)	Output power (W)	Selected motor
Joint 1	87.1	4.52	393.69	MG996R servo
Joint 2	72.94	8.11	591.54	MG996R servo
Joint 3	10.28	1.01	10.38	MG996R servo
Joint 4	29.86	3.30	98.54	SG90 Micro servo
Joint 5	10.10	1.10	11.11	SG90 Micro servo
Joint 6 (base link)	15.82	0.77	12.18	MG996R servo

```
import ETS3.*

dh = [
     0 0.287 0 1.571
     0 0 0.328 0
     0 0 0 1.571
     0 0.318 0 1.571
     0 0 0 -1.571
     0 0.253 0 0
     ]
r = SerialLink(dh, 'name', 'IRM')

qz = [0,0,0,0,0,0]
qr = [0,pi/9, pi/2,0,pi/12,0]
qs = [0,0,-pi/2,0,0,0]
qn = [0, pi/20, 2*pi/3, 0, pi/6,0]

Tz = r.fkine(qz)
Tr = r.fkine(qr)
Ts = r.fkine(qs)
Tn = r.fkine(qn)

%r.plot(qn)

r.plot(qz)
r.teach(qz)
qi = r.ikine(Tn, 'ld')

J = r.jacob0(qn)
J1 = r.jacob0(qr)
inJ = inv(J)

jsingu(J)
jsingu(J1)

determinant_J = det(J)
determinant_J1 = det(J1)

qd = inv(J)*[0.5 0.5 0.5 0.3 0.3 0.3]'
qd'

qd = inv(J1)*[1 1 1 0.5 0.5 0.5]'
qd'
```

```
%r.vellipse(qn, 'trans')

r.maniplty(qn)

force_tau = r.jacob0(qn)' * [0 20 0 0 0 0]'
```

7.6 Recommendations and conclusion

MATLAB robotic toolbox provides a variety of essential tools required for robotic modeling, simulation, and analyzing experimental results. Compared to the direct kinematics, inverse kinematics is more complex since it consists of many analytical solutions. The DH parameters and transformation matrices were used for modeling the general forward and inverse kinematics of the manipulator. It took a long time to complete the inverse kinematics and a lot of errors were incurred. MATLAB robotic toolbox created by Peter Corke has been employed to validate the general solution and obtain the numerical solution. Due to the complexity of the equations, during execution a delay time of 2–3 min, depending on the processing speed of the computer and the spatial location of the point being reached for, has been encountered. Since the aim of the study is to validate the general solution, a trial and error method was used to achieve the final results. Dynamic modeling was done using the symbolic functions of the MATLAB toolbox. From the observations, it was seen that the toolbox is very effective, especially for higher degree of freedom manipulators, as it saves time and eliminates computational errors. The numerical solution obtained for the inverse dynamics using the Simulink model had output wave forms that needed tuning of the parameters. However, this was not a problem since some values were randomly selected such as the damping coefficient, just to obtain the desired result (torques). The dynamic numerical solution was now used to aid in motor sizing and selection process. MG996R and SG90 micro servo motors were selected to actuate the robotic manipulator joint, with the smaller servo being used to actuate spherical wrist of the arm. The research results will now to be used in development of the robotic arms which will be used to test COVID-19 in hospitals. This study is part of a design project, which is aimed at designing an intelligent robotic machine for testing COVID-19. Finally, for future work, we recommend an easier interconnection between the CAD and MATLAB software to be developed as we encountered some difficulties in doing so during this study.

References

Anne, T., Wilkinson, J., & Li, Z. (2021). *Meta-reinforcement learning for adaptive motor control in changing robot dynamics and environments.* Available at: http://arxiv.org/abs/2101.07599 (Accessed 23 April 2021).

Ansarieshlaghi, F., & Eberhard, P. (2019). Trajectory tracking control of a very flexible robot using a feedback linearization controller and a nonlinear observer. In *CISM international centre for mechanical sciences, courses and lectures* (pp. 26–33). Springer International Publishing. https://doi.org/10.1007/978-3-319-78963-7_5.

Balafoutis, C. A., & Patel, R. V. (1991). Manipulator inverse dynamics. In *Dynamic analysis of robot manipulators* (pp. 117–182). Springer US. https://doi.org/10.1007/978-1-4615-3952-0_5.

Bamford, A. (2006). *The WOW factor: Global research compendium on the impact of arts in education.* New York: Waxmann Munster. Springer International Publishing. https://doi.org/10.1007/978-3-319-78963-7_22.

Baressi Šegota, S., et al. (2020). Path planning optimization of six-degree-of-freedom robotic manipulators using evolutionary algorithms. *International Journal of Advanced Robotic Systems, 17*(2). https://doi.org/10.1177/1729881420908076. SAGE Publications Inc.

Barrientos, J., et al. (2021). Real-time kinematics of continuum robots: Modelling and validation MiRoR (miniaturised robotic systems for holistic in-situ repair and maintenance works in restrained and hazardous environments) view project through-life performance: From science to instrumentation. *Robotics and Computer-Integrated Manufacturing.* https://doi.org/10.1016/j.rcim.2020.102019. Elsevier.

Ben Hamida, I., et al. (2019). Geometric based approach for workspace analysis of translational parallel robots. In *CISM international centre for mechanical sciences, courses and lectures* (pp. 180–188). Springer International Publishing. https://doi.org/10.1007/978-3-319-78963-7_24.

Bencomo, S. D. (2002). Control learning: Present and future. *IFAC Proceedings Volumes, 15*(1), 71–93. https://doi.org/10.3182/20020721-6-es-1901.01641 (IFAC-PapersOnline).

Bhattu, A. A., & Kulkarni, S. (2021). Dynamics of tendon actuated continuum robots by Cosserat rod theory. In *CISM international centre for mechanical sciences, courses and lectures* (pp. 414–422). Springer Science and Business Media Deutschland GmbH. https://doi.org/10.1007/978-3-030-58380-4_50.

Boby, R. (2021). *Combination of geometric and parametric approaches for kinematic identification of an industrial robot.* Elsevier. Available at: https://www.sciencedirect.com/science/article/pii/S0736584521000272 (Accessed 23 April 2021).

Brinker, J., et al. (2019). Dynamic modeling of functionally extended delta-like parallel robots with virtual tree structures. In *CISM international centre for mechanical sciences, courses and lectures* (pp. 171–179). Springer International Publishing. https://doi.org/10.1007/978-3-319-78963-7_23.

Chablat, D., Venkateswaran, S., & Boyer, F. (2019). Dynamic model of a bio-inspired robot for piping inspection. In *CISM international centre for mechanical sciences, courses and lectures* (pp. 42–51). Springer International Publishing. https://doi.org/10.1007/978-3-319-78963-7_7.

Chen, A. S., et al. (2021). Batch exploration with examples for scalable robotic reinforcement learning. *IEEE Robotics and Automation Letters, 6*(3), 4401–4408. https://doi.org/10.1109/LRA.2021.3068655. Institute of Electrical and Electronics Engineers Inc.

DeBoon, B. (2021). *Nine degree-of-freedom kinematic modeling of the upper-limb complex for constrained workspace evaluation.* asmedigitalcollection.asme.org. Available at: https://asmedigitalcollection.asme.org/biomechanical/article/143/2/021009/1087284 (Accessed 23 April 2021).

Dong, H., et al. (2019). Kinematic model of harmonic drive in robot joints with input eccentricity error. In *CISM international centre for mechanical sciences, courses and lectures* (pp. 134–140). Springer International Publishing. https://doi.org/10.1007/978-3-319-78963-7_18.

Featherstone, R. (2007). *Robot dynamics*. Scholarpedia. Available at: http://www.scholarpedia.org/article/Robot_dynamics (Accessed 24 July 2020).

Ferrentino, E., & Chiacchio, P. (2019a). A topological approach to globally-optimal redundancy resolution with dynamic programming. In *CISM international centre for mechanical sciences, courses and lectures* (pp. 77–85). Springer International Publishing. https://doi.org/10.1007/978-3-319-78963-7_11.

Ferrentino, E., & Chiacchio, P. (2019b). Topological analysis of global inverse kinematic solutions for redundant manipulators. In *CISM international centre for mechanical sciences, courses and lectures* (pp. 69–76). Springer International Publishing. https://doi.org/10.1007/978-3-319-78963-7_10.

Fomin, A., Glazunov, V., & Terekhova, A. (2019). Development of a novel rotary hexapod with single drive. In *CISM international centre for mechanical sciences, courses and lectures* (pp. 141–146). Springer International Publishing. https://doi.org/10.1007/978-3-319-78963-7_19.

Gil, A., et al. (2015). Development and deployment of a new robotics toolbox for education. *Computer Applications in Engineering Education, 23*(3), 443–454. https://doi.org/10.1002/cae.21615.

Gong, Z., et al. (2021a). A soft manipulator for efficient delicate grasping in shallow water: Modeling, control, and real-world experiments. *International Journal of Robotics Research, 40*(1), 449–469. https://doi.org/10.1177/0278364920917203. SAGE Publications Inc.

Gong, Z., et al. (2021b). A soft manipulator for efficient delicate grasping in shallow water: Modeling, control, and real-world experiments. *The International Journal of Robotics Research*. journals.sagepub.com. Available at: https://journals.sagepub.com/doi/abs/10.1177/0278364920917203 (Accessed 23 April 2021).

González, F. A. D., Cabrera, P. R. P., & Calderón, C. M. H. (2016). Design of a nanosatellite ground monitoring and control software—A case study. *Journal of Aerospace Technology and Management, 8*(2), 211–231. https://doi.org/10.5028/jatm.v8i2.553.

Guan, D., et al. (2021). Kinematic modeling and constraint analysis for robotic excavator operations in piling construction. *Automation in Construction*. Elsevier. Available at: https://www.sciencedirect.com/science/article/pii/S0926580521001175 (Accessed 23 April 2021).

Guo, W., et al. (2016). Kinematics, dynamics, and control system of a new 5-degree-of-freedom hybrid robot manipulator. *Advances in Mechanical Engineering, 8*(11), 1–19. https://doi.org/10.1177/1687814016680309. SAGE Publications Inc.

Herias Candelas, A. F., & Jara, A. C. (2006). *(PDF) Flexible virtual and remote laboratory for teaching robotics*. Available at: https://www.researchgate.net/publication/39438090_Flexible_virtual_and_remote_laboratory_for_teaching_Robotics (Accessed 23 April 2021).

Hill, R. B., et al. (2019). Minimizing input torques of a high-speed five-bar mechanism by using variable stiffness springs. In *CISM international centre for mechanical sciences, courses and lectures* (pp. 61–68). Springer International Publishing. https://doi.org/10.1007/978-3-319-78963-7_9.

Hu, Y., et al. (2019). The design of legged mobile welding robot. In *CISM international centre for mechanical sciences, courses and lectures* (pp. 102–109). Springer International Publishing. https://doi.org/10.1007/978-3-319-78963-7_14.

Huang, X., & Zou, J. (2021). *Kinematic modeling and control of variable curvature soft continuum robots*. ieeexplore.ieee.org. Available at: https://ieeexplore.ieee.org/abstract/document/9340011/ (Accessed 23 April 2021).

Hwang, Y.-L. (2006). Recursive Newton-Euler formulation for flexible dynamic manufacturing analysis of open-loop robotic systems. *The International Journal of Advanced Manufacturing Technology*, *29*(5), 598–604. https://doi.org/10.1007/s00170-005-2530-3. Springer.

Jin, J. (2021a). A robust zeroing neural network for solving dynamic nonlinear equations and its application to kinematic control of mobile manipulator. *Complex & Intelligent Systems*, 7(1), 87–99. https://doi.org/10.1007/s40747-020-00178-9. Springer Science and Business Media LLC.

Jin, J. (2021b). A robust zeroing neural network for solving dynamic nonlinear equations and its application to kinematic control of mobile manipulator. *Complex & Intelligent Systems*. Springer. Available at: https://link.springer.com/article/10.1007/s40747-020-00178-9?error=cookies_not_supported&error=cookies_not_supported&code=9c690de1-ea4f-4b6d-8c69-88267c667085&code=266ecf7b-1d15-4062-ab08-ee45d4307f5c (Accessed 23 April 2021).

Kaur, U., et al. (2021). Collision detection and inverse dynamics control of KUKA LBR IIWA robot. *International Journal of Mechatronics and Automation*, *8*(1), 9–21. https://doi.org/10.1504/IJMA.2021.113718. Inderscience Publishers.

Khalil, W. (2011). Dynamic modeling of robots using Newton-Euler formulation. In *Lecture notes in electrical engineering* (pp. 3–20). https://doi.org/10.1007/978-3-642-19539-6_1.

Kim, J. Y., & Kim, J. Y. (2021). Gait training algorithm based on inverse dynamics of walking rehabilitation robot, DDgo Pro. *Intelligent Service Robotics*. https://doi.org/10.1007/s11370-021-00357-8. Springer Science and Business Media Deutschland GmbH.

Lin, R., & Guo, W. (2019). Type synthesis of a legged mobile lander based on a given truss. In *CISM international centre for mechanical sciences, courses and lectures* (pp. 118–125). Springer International Publishing. https://doi.org/10.1007/978-3-319-78963-7_16.

Osgouie, K. G., & Gard, B. (2017). Using the matrix method to compute the degrees of freedom of mechanisms. *Journal of Applied and Computational Mechanics*, *3*(3), 158–170. https://doi.org/10.22055/jacm.2017.20542.1085.

Otani, T., et al. (2019). Ankle and foot mechanism mimicking joint stiffness and following motion based on human. In *CISM international centre for mechanical sciences, courses and lectures* (pp. 86–93). Springer International Publishing. https://doi.org/10.1007/978-3-319-78963-7_12.

Owen-Hill, A. (2015). *Kinematics: Why robots move like they do*. Available at: https://blog.robotiq.com/kinematics-why-robots-move-like-they-do (Accessed 24 July 2020).

Peidró, A., et al. (2019). Parallelisms between planar and spatial tricept-like parallel robots. In *CISM international centre for mechanical sciences, courses and lectures* (pp. 155–162). Springer International Publishing. https://doi.org/10.1007/978-3-319-78963-7_21.

Penčić, M., et al. (2019). Robotic eyes with 7 DOFs: Structural design and motion simulation. In *CISM international centre for mechanical sciences, courses and lectures* (pp. 52–60). Springer International Publishing. https://doi.org/10.1007/978-3-319-78963-7_8.

Perrusquia, A., & Yu, W. (2021). *Multi-agent reinforcement learning for redundant robot control in task-space*. Springer. Available at: https://link.springer.com/article/10.1007/s13042-020-01167-7 (Accessed 23 April 2021).

Perrusquía, A., Yu, W., & Li, X. (2021). Multi-agent reinforcement learning for redundant robot control in task-space. *International Journal of Machine Learning and Cybernetics*, *12*(1), 231–241. https://doi.org/10.1007/s13042-020-01167-7. Springer Science and Business Media Deutschland GmbH.

Plancher, B., et al. (2021). Accelerating robot dynamics gradients on a CPU, GPU, and FPGA. *IEEE Robotics and Automation Letters*, *6*(2), 2335–2342. https://doi.org/10.1109/LRA.2021.3057845. Institute of Electrical and Electronics Engineers Inc.

Righetti, L., et al. (2011). *Inverse dynamics control of floating-base robots with external constraints: A unified view*. ieeexplore.ieee.org. Available at: https://ieeexplore.ieee.org/abstract/document/5980156/ (Accessed 24 July 2020).

Righettini, P., et al. (2015). Symbolic kinematic and dynamic modelling toolbox for multi-DOF robotic manipulators. In *2015 21st international conference on automation and computing: Automation, computing and manufacturing for new economic growth, ICAC 2015, (September)*. https://doi.org/10.1109/IConAC.2015.7313939.

Rodríguez León, J. F., et al. (2019). Experiences and design of a cable-driven assisting device for arm motion. In *CISM international centre for mechanical sciences, courses and lectures* (pp. 94–101). Springer International Publishing. https://doi.org/10.1007/978-3-319-78963-7_13.

Rouse, W. B. (2013). The conditions of complex innovation adoption occurence. *The Electric Journal Information System Evaluation, 16*, 10.

Sciavicco, L., Siciliano, B., & Villani, L. (1995). Lagrange and Newton-Euler dynamic modeling of a gear-driven robot manipulator with inclusion of motor inertia effects. *Advanced Robotics, 10*(3), 317–334. https://doi.org/10.1163/156855395X00427.

Serrezuela, R. R., et al. (2017). Kinematic modelling of a robotic arm manipulator using MATLAB. *ARPN Journal of Engineering and Applied Sciences, 12*(7), 2037–2045.

Shen, H., et al. (2021). Kinematic sensitivity, parameter identification and calibration of a non-fully symmetric parallel Delta robot. *Mechanism and Machine Theory*. Elsevier. Available at: https://www.sciencedirect.com/science/article/pii/S0094114X21000690 (Accessed 23 April 2021).

Shin, K. G., & Mckay, N. D. (1986). A dynamic programming approach to trajectory planning of robotic manipulators. *IEEE Transactions on Automatic Control, 31*(6), 491–500. https://doi.org/10.1109/TAC.1986.1104317.

Sun, Q., Angeles, J., & Forbes, J. R. (2019). Elastostatics of a full-mobility PKM with flexible links. In *CISM international centre for mechanical sciences, courses and lectures* (pp. 34–41). Springer International Publishing. https://doi.org/10.1007/978-3-319-78963-7_6.

Tan, H. H., & Potts, R. B. (1988). Minimum time trajectory planner for the discrete dynamic robot model with dynamic constraints. *IEEE Journal on Robotics and Automation, 4*(2), 174–185. https://doi.org/10.1109/56.2081.

Toussaint, M. (2014). *Robotics kinematics. The Robotics course (winter term 2014/15, U Stuttgart), including a bullet point list to help prepare for exams*. University of Stuttgart.

Tsai, L. (2000). *Solving the inverse dynamics of a Stewart-Gough manipulator by the principle of virtual work*. asmedigitalcollection.asme.org. Available at: https://asmedigitalcollection.asme.org/mechanicaldesign/article/122/1/3/443751 (Accessed 23 April 2021).

Wu, G., & Caro, S. (2019). Torsional stability of a U-joint based parallel wrist mechanism featuring infinite torsion. In *CISM international centre for mechanical sciences, courses and lectures* (pp. 147–154). Springer International Publishing. https://doi.org/10.1007/978-3-319-78963-7_20.

Xing, D., et al. (2021). *General robot dynamics learning and Gen2Real*. Available at: http://arxiv.org/abs/2104.02402 (Accessed 23 April 2021).

Zhang, D., & Peng, Z. (2021). *Positioning accuracy reliability of industrial robots through probability and evidence theories*. asmedigitalcollection.asme.org. Available at: https://asmedigitalcollection.asme.org/mechanicaldesign/article-abstract/143/1/011704/1084408 (Accessed 23 April 2021).

Further reading

Kindel Media. (2021). Full shot of robot toy. *Studies in Health Technology and Informatics*. https://doi.org/10.3233/978-1-60750-706-2-716.

CHAPTER 8

Design optimization of a medical robotic manipulator for testing COVID-19 virus using finite element analysis

8.1 Introduction

Medical robots consist of different types of technical components and machines used in the medical field either to perform mechanical work or to aid such exercises (Bendel, 2015; Rupal et al., 2017). Medical robots are reliable assistant that perform errands that go past the abilities of people. A significant factor in this regard is the beginning of exhaustion in people during extensive surgical operations. Driven by advancement and their incredible significance in society, the quantities of affirmed medical robots being used on a worldwide level is developing, empowering the field of biomedical engineering and other additional fields as well (Girone et al., 2001; van Wynsberghe, 2013).

The major modules of medical robots normally involve a controller unit and manipulator. Robotic manipulators are designed to perform activities and comprise of jointed fragments, sensors, a drive framework, and the effector (Hagn et al., 2008; Taylor et al., 2016). While electrical drive innovation is normally utilized, the quantity of fragments and joints relies upon a particular application (Alric et al., 2009; Taylor et al., 2016). The sensor framework is not just answerable for deciding the controller's position but it also gathers different kinds of information, for example, distances and camera pictures and feeds it to the robot's regulator.

The historical backdrop of medical robots returns around 25 years, which makes them a generally ongoing accomplishment of biomedical engineering (Challacombe et al., 2006). In any case, since their beginning, the field of mechanical technology in medication has immediately settled itself and is presently irreplaceable in numerous applications. The first time a robot was utilized to aid surgical operations was in 1991, when a transurethral

Healthcare Systems Design of Intelligent Testing Centers
https://doi.org/10.1016/B978-0-323-99443-9.00006-1

resection of the prostate (an interventional urological strategy for unhealthy tissue) was done; this was immediately trailed by several further methods (Challacombe et al., 2006; Ewing et al., 2004).

Medical robots fundamentally are for assisting health (Fatima et al., 2021; Nawrat, 2006). They provided a finely nuanced choice of uses and methods, and they are designed to perform specific processes with incredible precision and effortlessness (Wolf & Shoham, 2009). The principal benefits and elementary attributes of medical robots are brought to bear in rudimentary invasive methodology (Taylor et al., 2016). Their consistent degree of work execution serves to abbreviate the span of operations. This in turn lessens patient stress and diminishes the rate of treatment mistakes achieved through tiredness. The capacity of medical robots to work with extraordinary precision is of specific significance in minimum invasive surgery, which depends on the utilization of exceptionally small incisions (Taylor et al., 2016; Zhang & Lu, 2020). Such frameworks are equipped for downsizing actions performed by doctors and changing them to minimal activities (Goher, Mansouri, & Fadlallah, 2017).

Investigation in low-cost robotics can have various benefits in both speeding up human development by making robots more broadly adopted and making assistive robots more moderate for everybody (Hilder et al., 2016). Relative cheap robotic manipulators that are adroit, profoundly programmable, particular, and simple to assembly are especially wanted (Kalo & Newsum, 2014; Scott, 2020). Numerous robotic manipulators that can automate an assignment with accuracy ordinarily accompany an exorbitant price, due to high-precise actuators utilized in their joints along with uniquely machined metal links and advanced algorithms. Less expensive assembling and manufacturing methods, materials, and actuators can assist advance robotics technology innovation (Mavroidis, 2002; Polygerinos et al., 2017; Yap, Chen, & Yeow, 2016). This will enable a quicker reconciliation of robots in our everyday life.

8.2 Literature review

Roy, Ghosh, and Mishra (2018) conducted a research on dynamic finite element analysis and optimization of a robotic arm for CT image-guided procedures with the aim to design a robotic manipulator that will be relative cheap in the method of biopsy. Prime goal of this exploration was to decide the structural precision of the robotic manipulator in the long run. The virtual model of the robotic manipulator was developed in SolidWorks 2016.

A trial model of the robotic manipulator was made to check deviation in arriving at a client characterized target point. The design was exposed to finite element analysis in ANSYS 15.0. Findings of the study revealed that a net decrease in the maximum total deformation of 91.499% can be accomplished by making some necessary modifications. Therefore, with these alterations, the accuracy of this robotic manipulator can be maintained for a larger time frame and has better uses in the medical field.

A study was conducted by Khatami (2002) on isotropic design optimization of robotic arm utilizing a genetic algorithm procedure, where the kinematic isotropy was considered as the performance assessment basis for optimal development of robotic manipulators. A genetic algorithm was developed which was created to address the optimal optimization definition of robot design to track down the ideal design boundaries. The algorithm was executed to streamline through the entire robot's workspace.

A research was done on topology optimization of rigid links for industrial arm considering dynamic loading conditions by Srinivas (2020). The produced topology was not ideal for other precise positions, as it is reliant upon load direction instead of load magnitude. A strategy was proposed which is comparable to static procedure to integrate a solitary topology, which performs better in every angular position. Findings show that 10%–25% decrease in deflection and stress can be achieved by this method compared to other optimal topologies. The simulated results were approved utilizing experimentations. Furthermore, it was also shown that a decrease of 30% link volume reduces joint torques by 24.9%, with optimal values of deflection and stress.

Zhou et al. (2012) conducted a study on integrated dimensional and drive train design optimization of a lightweight anthropomorphic arm. The main objective was to minimize the mass of robotic manipulators. The strategy tends to the impacts of measurements and attributes of drive trains in the design improvement. Imperatives were formed based on kinematic execution and dynamic necessities, though the fundamental goal was to limit the total mass. Contextual investigations were incorporated to show the utilization of the optimization technique in the development of assistive robots.

A study on modal and harmonic response analysis of key components of robotic arm based on ANSYS was done by Tang, Yu, and Shi (2017). To investigate the problem, a three-dimensional virtual model was developed in the SolidWorks. The model was then imported into ANSYS for modal analysis. The earlier mentioned analysis was used to acquire the natural frequency of vibration and the impact of key robotic arm components. Therefore the machine can be operated such that its vibration frequency is not the same with the natural frequency of manipulator, and thus avoiding poor working conditions due to resonance. The harmonic response analysis was also done based on the modal analysis. This analysis enabled the determination of displacement and stress distribution diagrams of harmonic response. Using this analysis, a reference value was obtained which enables structural design optimization of the robotic manipulator.

8.3 Materials and methods

Topology optimization locates sections of a structure where material weight can be reduced, thereby enabling design optimization. SolidWorks 2018 has been used as the FEA tool. During topology optimization, supports and loads acting on a volume of material, optimization region, and constraints to meet the desired specifications must be specified. The medical robotic manipulator for testing COVID-19 virus was also modeled using SolidWorks 2018. Robotic manipulator components are made of Kevlar 53% and the material properties can be accessed at Oliwa (2020).

8.3.1 Structural loading

For this design analysis, only the robotic manipulator components were considered while the spherical wrist and end effector components were not. Several loading cases were employed for each individual components of the robotic manipulator. Each robotic manipulator component was analyzed based on the worst-case scenario. It was assumed that for the base link, the maximum load that can be exerted occurs when the arm is vertically stretched. For the link 1 subassembly 2, same loading conditions were assumed similar to those of the base link. Furthermore, link 1 subassembly 2 was assumed to carry a vertical load at the end equivalent to the total weight of links preceding it, plus the end effector. This is because link 1 subassembly 2 is also a component of link 2. Link 2 subassembly 1 and link 3 were also assumed to be loaded the same way as link 2 was loaded for the second time. Loading due to twisting moment was not considered for all of the links. It was assumed that vertical loading will be enough for the design optimization since large safety of factors were used (the components were initially overdesigned). Note that, large safety of factor is essential for a robotic manipulator that will be interacting directly with people, that is, to

Table 8.1 The force acting on each link.

Part	Force (N)	Factor of safety	Crictical load (N)
Base link	49.8	2	99.6
Link 1	45.3	2	90.6
Link 2	43.5	2	87
Link 3	34	2	68
Link 4	29.3	2	58.6
Link 5	23	2	46
End effector	21.4	2	42.8

enhance safety. Loads acting on each link of the robotic manipulator are summarized in Table 8.1.

8.3.2 Static structural analysis

To obtain the displacements, stresses, and strains acting on the robotic manipulator components, finite element analysis structural analysis was performed. Static simulation tool in SolidWorks 2018 was used to perform the static structural analysis. The Von Mises stress failure criterion was used for the design. Kevlar 53% is the material that was used for the robotic manipulator with a yield strength of 2758 MPa. To improve the solution accuracy, uniform fine mesh was developed in SolidWorks 2018. The analysis was done using loading conditions summarized in the earlier table. For the base link was taken as the fixed support and a force of 100 N applied vertically downward where the servo motor will be mounted. For the link 1 subassembly 2, a vertical force of 91 N was applied in the negative x-axis direction where the servo motor will be mounted and it was fixed at its end. The same procedure was used for link 2 subassembly 1 and link 3, using the loading conditions summarized in the earlier table. As shown in earlier table, a safety factor of 2 has been used to cater for the weight of motors, connecting cables, and sensors which were not included during load calculations. Also, it has been assumed that the maximum load that can be carried by the end effector equals 2 kg. The fixed supports for each robotic component are represented by green arrows and the loads are applied at the other ends (see Fig. 8.1).

8.3.3 Modal analysis

Modal analysis of the robotic manipulator components was performed to obtain the natural frequencies and mode shapes. SolidWorks 2018 frequency response simulation tool was used for modal analysis of the robotic

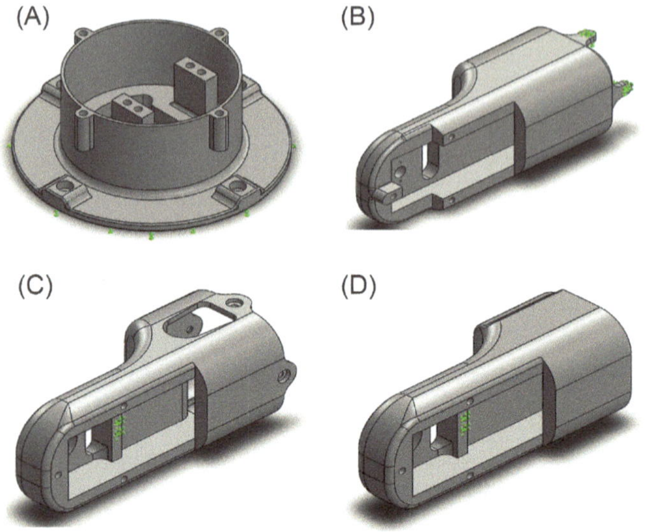

Fig. 8.1 Support points of the base link (A), link 1 subassembly 2 (B) Support points of link 2 subassembly 1(C) and link 3 (D). (Data from *Marvellous Moyo and Tawanda Mushiri.*)

manipulator components. Each component was analyzed with only fixed fixtures at their respective supports (fixed supports are the same as those discussed before) and no external loads were applied. Only five mode shapes were done and the corresponding frequencies were obtained. These frequencies were then compared with the excitation frequency of the robotic manipulator which can be assumed to be 50 Hz according to JSUMO (2021). To avoid resonance the natural frequency of each individual component of the robotic manipulator should not match the excitation frequency.

8.3.4 Topology optimization

Topology optimization is a recent numerical procedure utilized to come up with an optimum shape and material distribution for a given model under preset design constraints. Results obtained from topology optimization are not precisely accurate, but they give a rudimentary framework of the improved design solution. For the current robotic manipulator components, the optimization objective was structural weight reduction with the constraint being the percentage reduction of mass. This was achieved using SolidWorks 2018 topology simulation tool. Various percentages of weight reduction were considered in order to make sure that adequate material was left to preserve bolt locations while decreasing the original robotic components' weight. Prior to arriving at the final design, other factors were considered, such as ease of manufacturability and assembly and safety.

8.4 Results

8.4.1 Stress and displacement analysis

Structural static analysis was done to determine the total displacement and corresponding stresses induced in each robotic manipulator component. These results are categorized into individual sections corresponding to each robotic component part and these are summarized as follows.

8.4.1.1 Base link static structural analysis results

Figs. 8.2 and 8.4 illustrate original base link results due to static loading, while Figs. 8.3 and 8.5 show the same solution but for the optimized link. From the earlier mentioned simulation results, maximum displacement of 2.241×10^{-4} mm occurs at the top of the servo motor mountings for the original base link, as shown in Fig. 8.2. For the optimized base link, a maximum displacement of 2.242×10^{-4} mm is achieved at the same location as that of the original link (see Fig. 8.3).

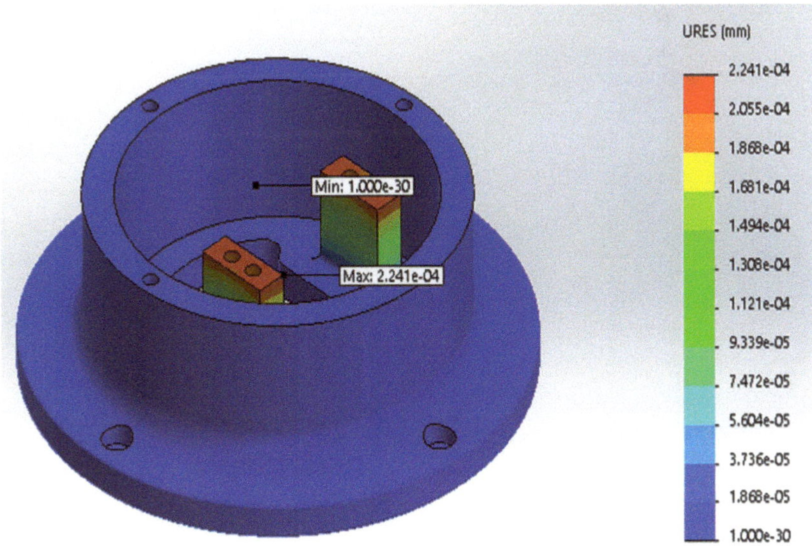

Fig. 8.2 Total deformation of the original base link. *(Data from Marvellous Moyo and Tawanda Mushiri.)*

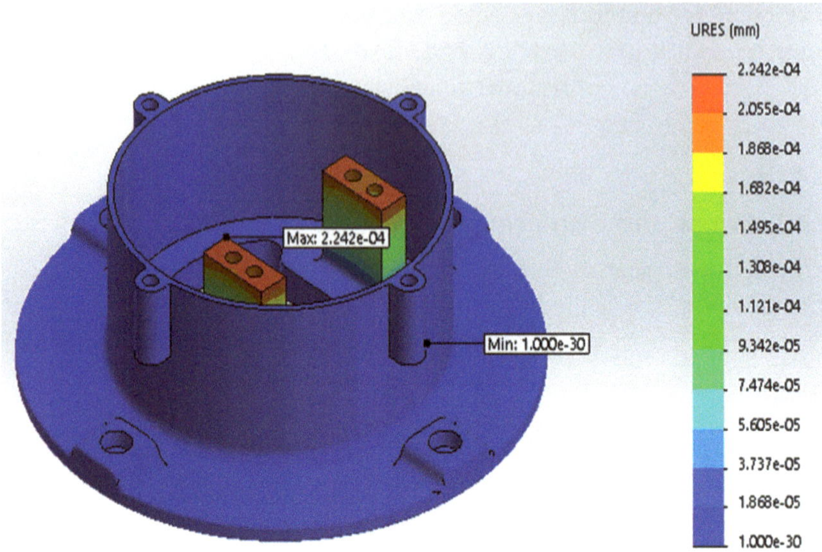

Fig. 8.3 Total deformation of the optimized base link. *(Data from Marvellous Moyo and Tawanda Mushiri.)*

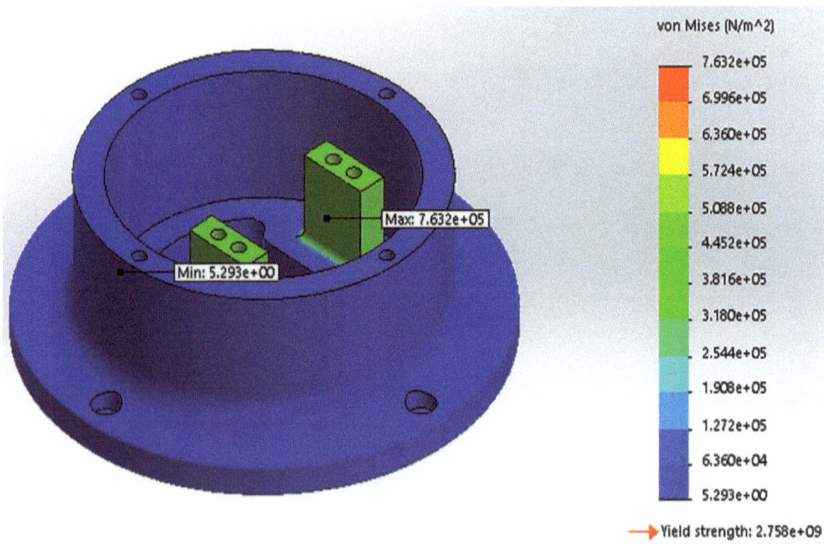

Fig. 8.4 Stress distribution of the original base link. *(Data from Marvellous Moyo and Tawanda Mushiri.)*

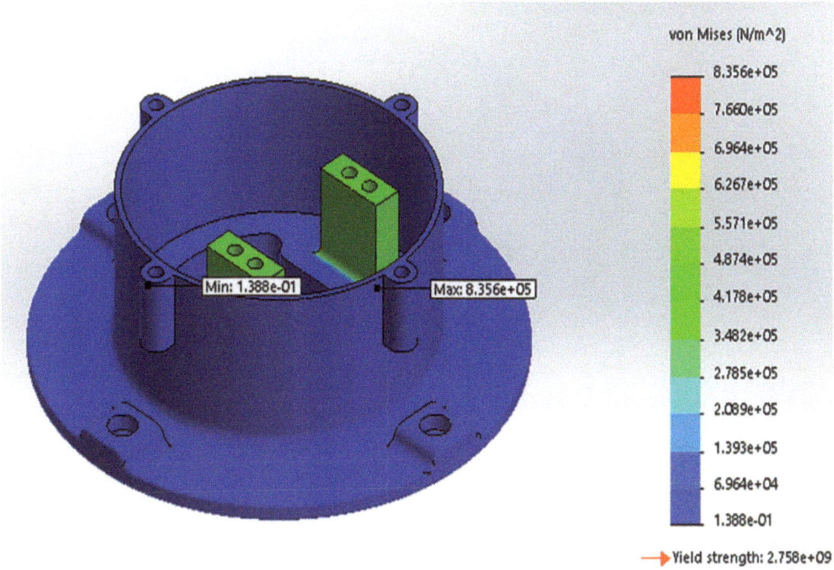

Fig. 8.5 Stress distribution of the optimized base link. *(Data from Marvellous Moyo and Tawanda Mushiri.)*

The Von Mises analysis stress results show that a maximum stress of 0.7632 MPa occurs on the original base link, as shown in Fig. 8.4 while a maximum stress of 0.8356 MPa occurs on the optimized base link (see Fig. 8.5). In both cases, these stresses are experienced on servo motor mounting sides.

8.4.1.2 Link 1 subassembly 2 static structural analysis results

Figs. 8.6 and 8.8 illustrate link 1 subassembly 2 results due to static loading, while Figs. 8.7 and 8.9 show the same solution but for the optimized link. From the earlier mentioned simulation results, maximum displacement of 1.481×10^{-3} mm occurs near the servo motor mountings for the original base link, as shown in Fig. 8.6. For the optimized link, a maximum displacement of 2.632×10^{-3} mm is achieved at the same location as that of the original link (see Fig. 8.7). In this case the component will be part of link 2.

The Von Mises analysis stress results show that a maximum stress of 2.047 MPa occurs on the original link, as shown in Fig. 8.8 while a maximum stress of 2.005 MPa occurs on the optimized link, as shown in Fig. 8.9, when the component is part of link 1. In both cases, these stresses are experienced on servo motor mountings.

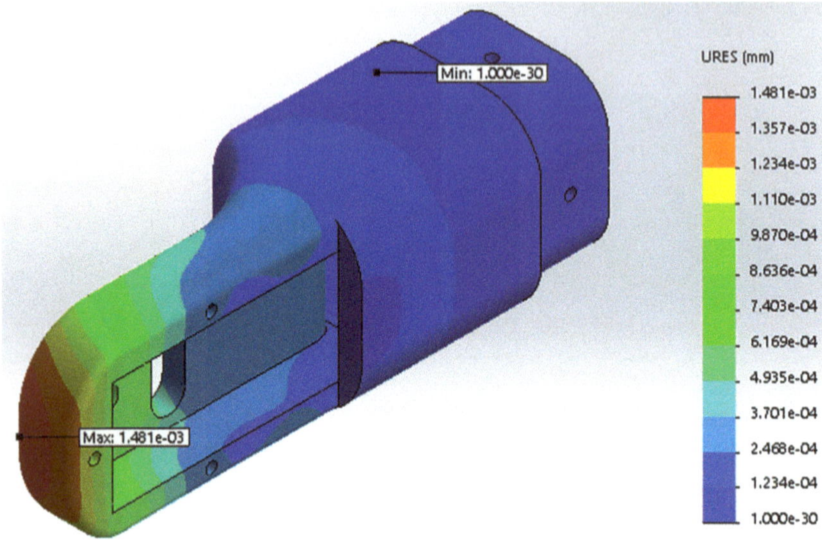

Fig. 8.6 Total deformation of the original link 1 subassembly 2. *(Data from Marvellous Moyo and Tawanda Mushiri.)*

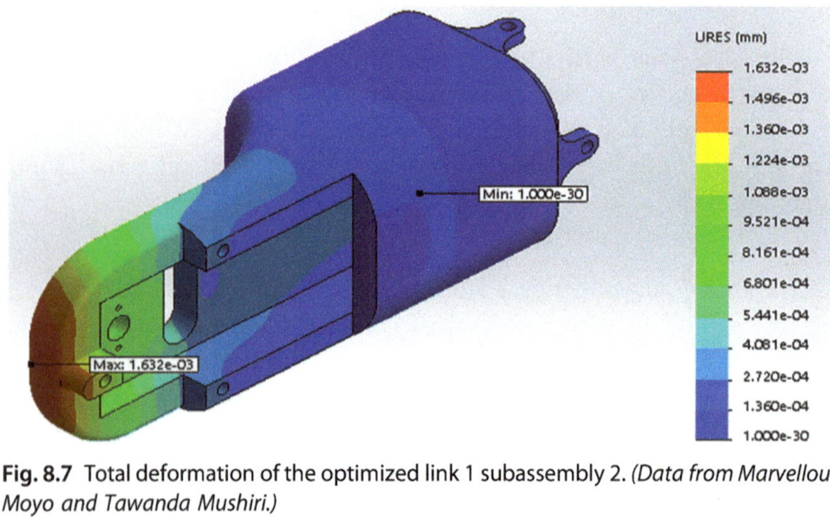

Fig. 8.7 Total deformation of the optimized link 1 subassembly 2. *(Data from Marvellous Moyo and Tawanda Mushiri.)*

Also, the simulation results show that a maximum displacement of 1.7×10^{-2} mm occurs near the servo motor mountings for the original link, as shown in Fig. 8.10. For the optimized link, a maximum displacement of 2.965×10^{-2} mm is achieved at the same location as that of the original link (see Fig. 8.11). In this case the component will be part of link 2.

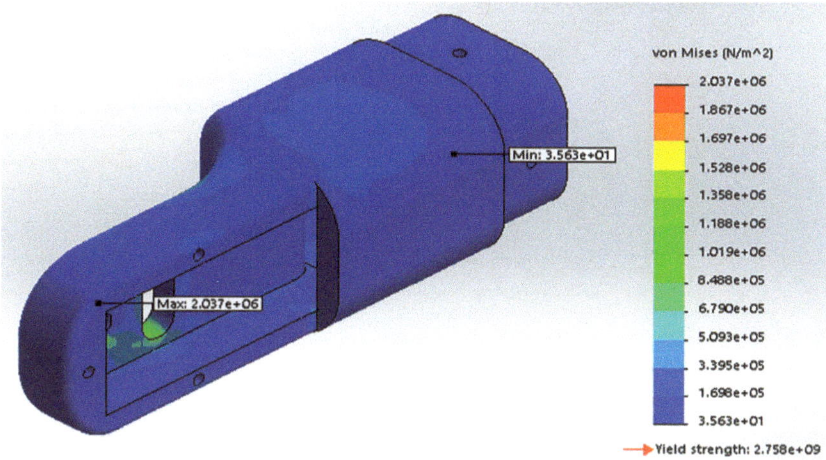

Fig. 8.8 Stress distribution of the original link 1 subassembly 2. *(Data from Marvellous Moyo and Tawanda Mushiri.)*

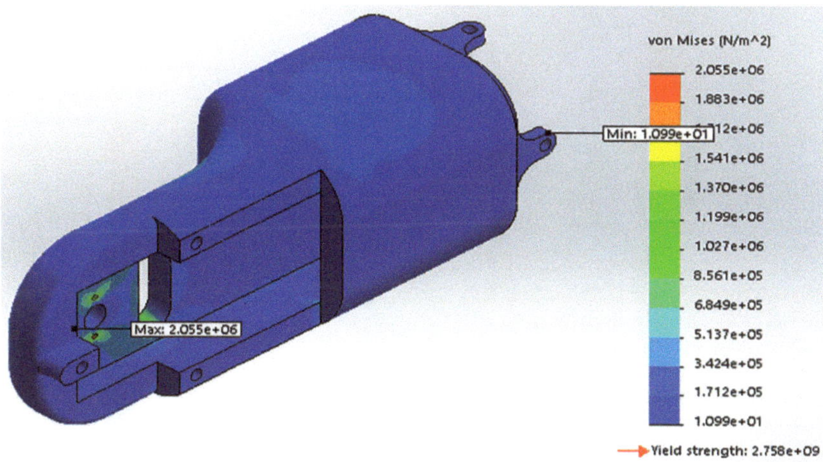

Fig. 8.9 Stress distribution of the optimized link 1 subassembly 2. *(Data from Marvellous Moyo and Tawanda Mushiri.)*

Further, the Von Mises analysis stress results show that a maximum stress of 4.523 MPa occurs on the original link, as shown in Fig. 8.12 while a maximum stress of 13.04 MPa occurs on the optimized link, as shown in Fig. 8.13, when the component is now part of link 2. In both cases, these stresses are experienced near the bolting holes, where link is fixed.

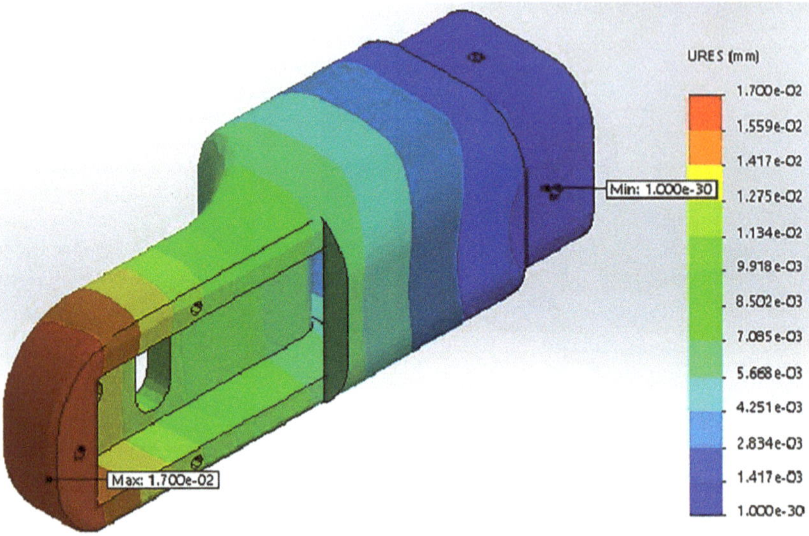

Fig. 8.10 Total deformation of the original link 2 subassembly 2. *(Data from Marvellous Moyo and Tawanda Mushiri.)*

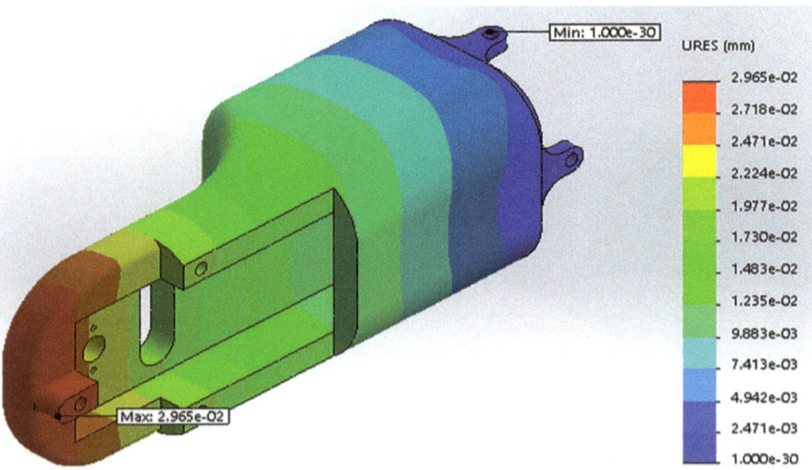

Fig. 8.11 Total deformation of the optimized link 2 subassembly 2. *(Data from Marvellous Moyo and Tawanda Mushiri.)*

8.4.1.3 Link 2 subassembly 1 static structural analysis results

Figs. 8.14 and 8.16 illustrate original link 2 subassembly 1 results due to static loading, while Figs. 8.15 and 8.17 show the same solution but for the optimized link. From the earlier mentioned simulation results, maximum displacement of 6.656×10^{-2} mm occurs near the bolting holes of the link, as shown in Fig. 8.14. For the optimized link, a maximum

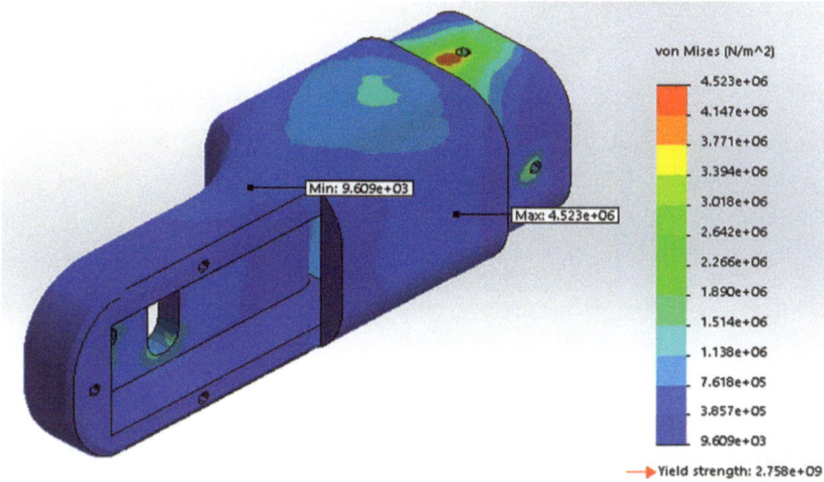

Fig. 8.12 Stress distribution of the original link 2 subassembly 2. *(Data from Marvellous Moyo and Tawanda Mushiri.)*

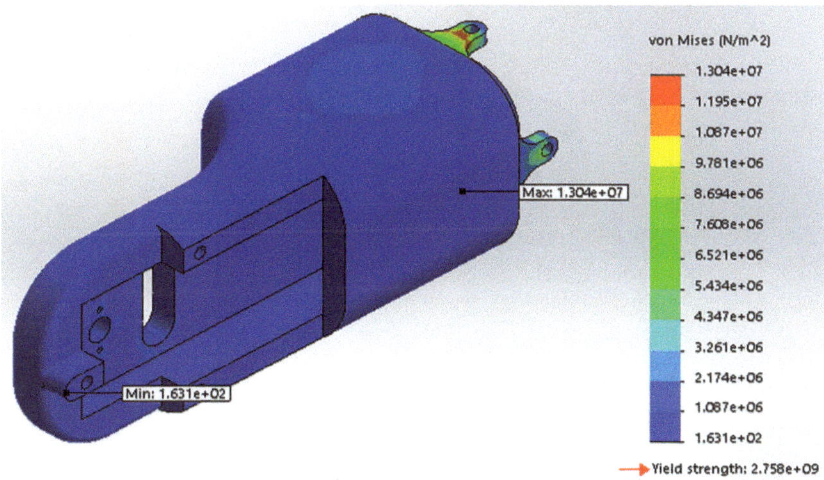

Fig. 8.13 Stress distribution of the optimized link 2 subassembly 2. *(Data from Marvellous Moyo and Tawanda Mushiri.)*

displacement of 1.236×10^{-1} mm is achieved at the same location as that of the original link (see Fig. 8.15).

The Von Mises analysis stress results show that a maximum stress of 47.51 MPa occurs on the original link, as shown in Fig. 8.16 while a maximum stress of 58.69 MPa occurs on the optimized link, as shown in Fig. 8.17. In both cases, these stresses are experienced around the bolting holes of the link.

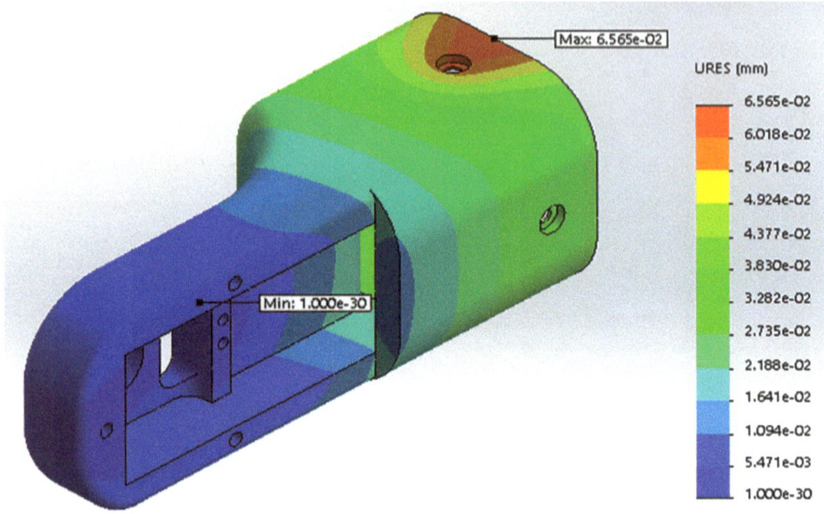

Fig. 8.14 Total deformation of the original link 2 subassembly 1. *(Data from Marvellous Moyo and Tawanda Mushiri.)*

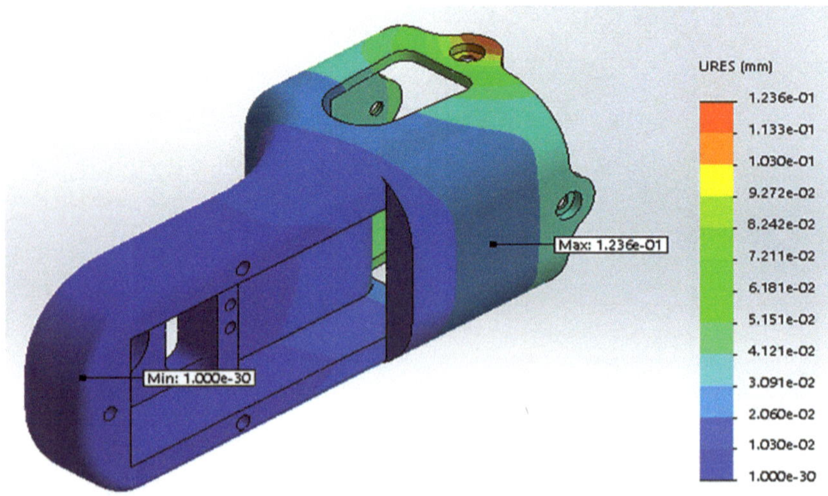

Fig. 8.15 Total deformation of the optimized link 2 subassembly 1. *(Data from Marvellous Moyo and Tawanda Mushiri.)*

8.4.1.4 Link 3 static structural analysis results

Figs. 8.18 and 8.20 illustrate original link 3 results due to static loading, while Figs. 8.19 and 8.21 show the same solution but for the optimized link. From the earlier mentioned simulation results, maximum displacement of 1.677×10^{-2} mm occurs near the servo motor mountings of the link, as

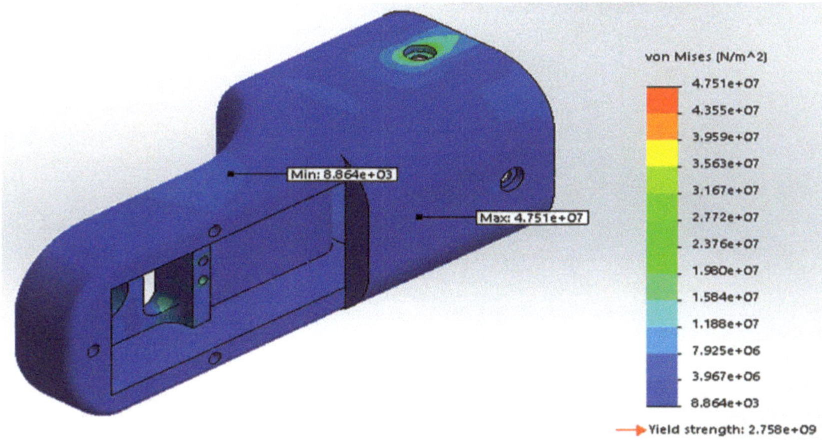

Fig. 8.16 Stress distribution of the original link 2 subassembly 1. *(Data from Marvellous Moyo and Tawanda Mushiri.)*

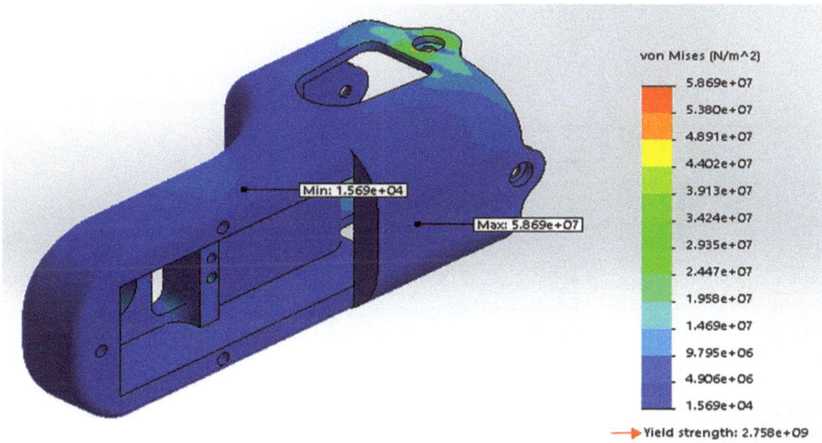

Fig. 8.17 Stress distribution of the optimized link 2 subassembly 1. *(Data from Marvellous Moyo and Tawanda Mushiri.)*

shown in Fig. 8.18. For the optimized link, the same maximum displacement is achieved at the same location as that of the original link (see Fig. 8.19).

The Von Mises analysis stress results show that a maximum stress of 14.07 MPa occurs on the original link, as shown in Fig. 8.20 while a maximum stress of 14.13 MPa occurs on the optimized link, as shown in Fig. 8.21. In both cases, these stresses are experienced around servo motor mountings of the link.

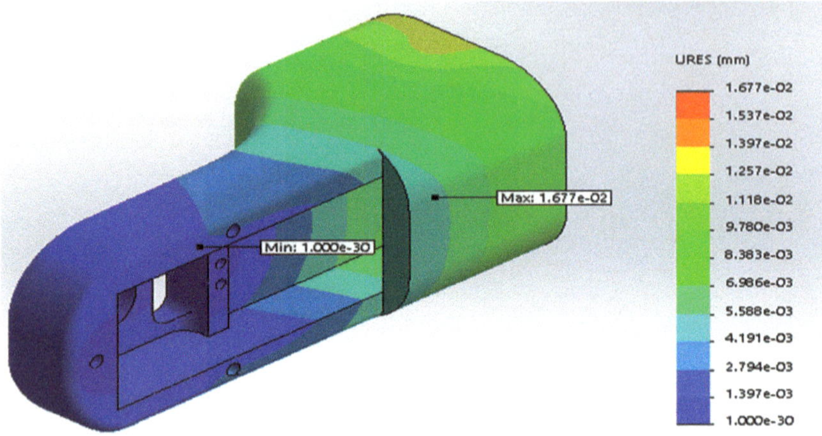

Fig. 8.18 Total deformation of the original link 3. *(Data from Marvellous Moyo and Tawanda Mushiri.)*

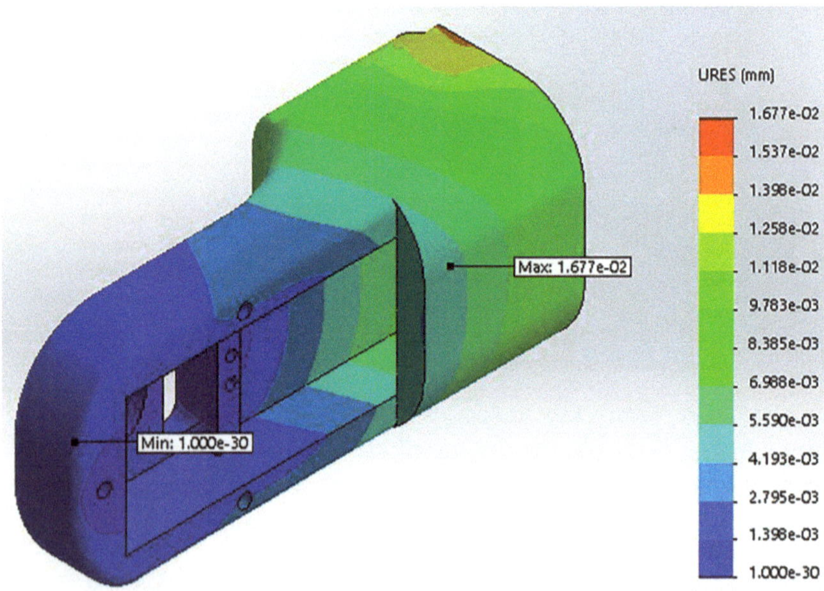

Fig. 8.19 Total deformation of the optimized link 3. *(Data from Marvellous Moyo and Tawanda Mushiri.)*

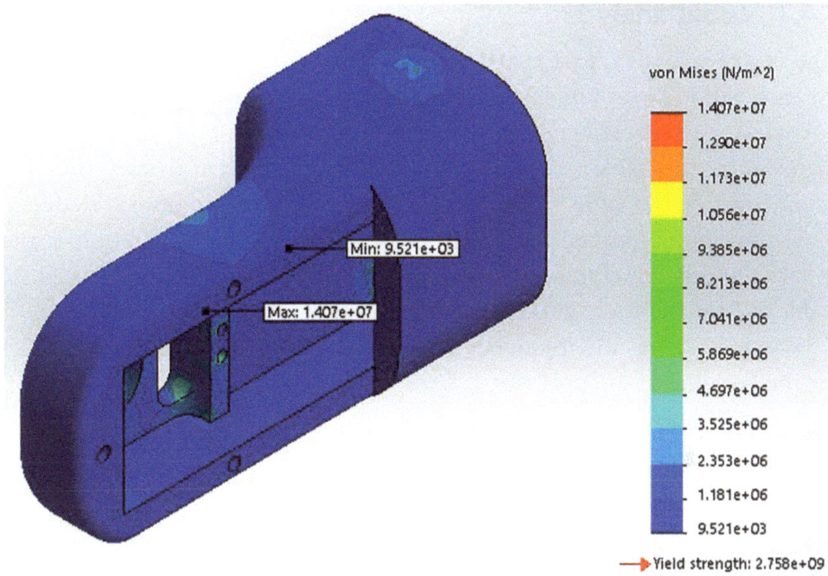

Fig. 8.20 Stress distribution of the original link 3. *(Data from Marvellous Moyo and Tawanda Mushiri.)*

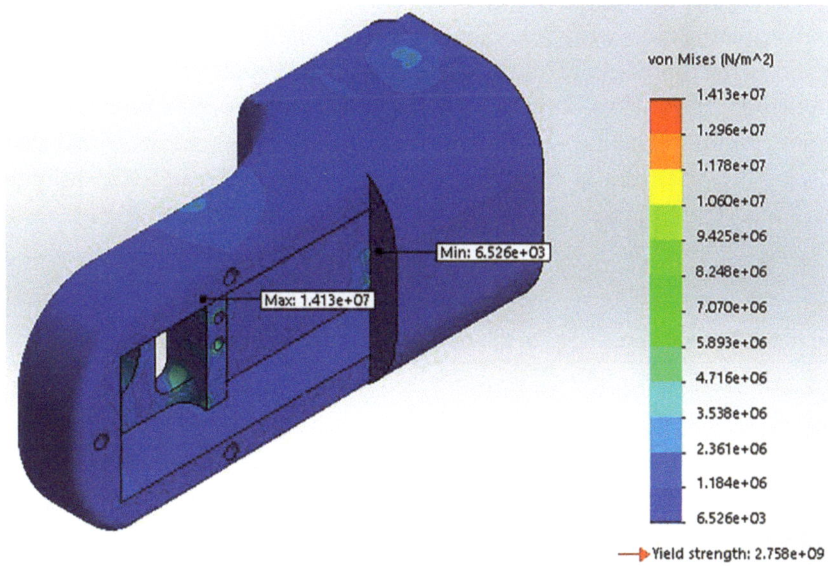

Fig. 8.21 Stress distribution of the optimized link 3. *(Data from Marvellous Moyo and Tawanda Mushiri.)*

8.4.2 Modal analysis

The modal analysis is performed on both the original and optimized design for each robotic manipulator components. The five modes (vibrational frequencies) are extracted and the results for each corresponding link are summarized in the table, as shown as follows. The frequencies that are summarized in the table were done under nonpreload conditions, to determine the natural frequencies that must be avoided during operation of the robotic testing machine to avoid resonance. Results are presented in Table 8.2 for the four links that were analyzed.

Fig. 8.22 shows the frequency response graph against number of modes for the base link.

Fig. 8.23 shows the frequency response graph against number of modes for link 1 subassembly 2.

Fig. 8.24 shows the frequency response graph against number of modes for link 2 subassembly 1.

Fig. 8.25 shows the frequency response graph against number of modes for link 3.

8.4.3 Topology optimization

A topology optimization was done for each link of the robotic manipulator under investigation with the objective of reducing the total mass. Since the results of the topology are just rough geometric proposals, geometrical interpretation was done to come up with the final design. To get a better view of the region of criticality, several iterations (using a slide tool in SolidWorks 2018) were performed to locate the regions where material density is required. The optimization results for each link are shown in Fig. 8.26.

Table 8.2 Natural frequencies of the optimized robotic manipulator components.

	Frequency (Hz)				
	Mode				
Part	**1**	**2**	**3**	**4**	**5**
Base link	5271.4	5642.1	5643.1	6247.2	6875.8
Link 1 subassembly 2	958.48	983.68	2972.3	3024.1	4331.1
Link 2 subassembly 1	1243.1	1903.6	2399.3	3818.4	5051.4
Link 3	1079.7	1838.8	1973.80	5987.8	6191.2

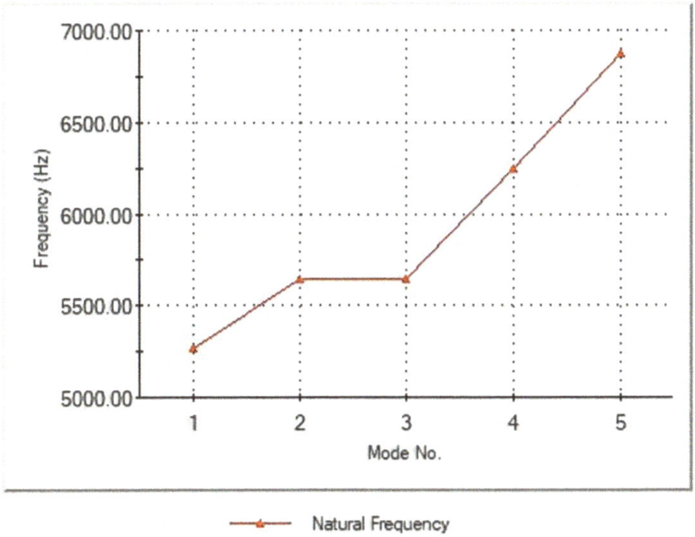

Fig. 8.22 Frequency response of the optimized base link. *(Data from Marvellous Moyo and Tawanda Mushiri.)*

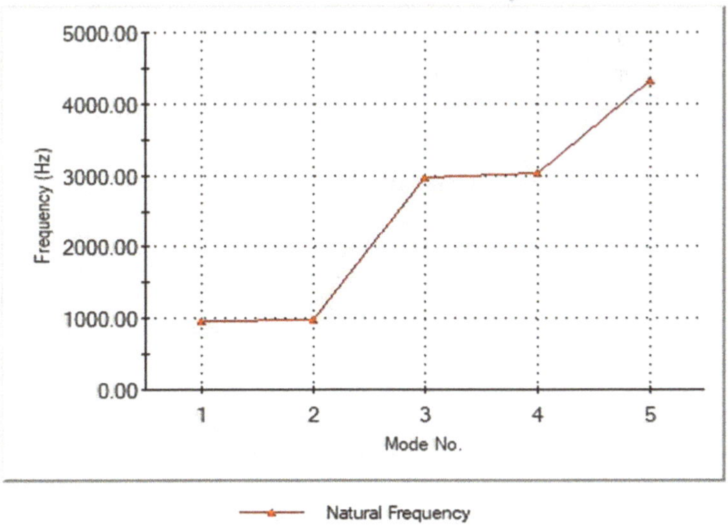

Fig. 8.23 Frequency response of the optimized link 1 subassembly 2. *(Data from Marvellous Moyo and Tawanda Mushiri.)*

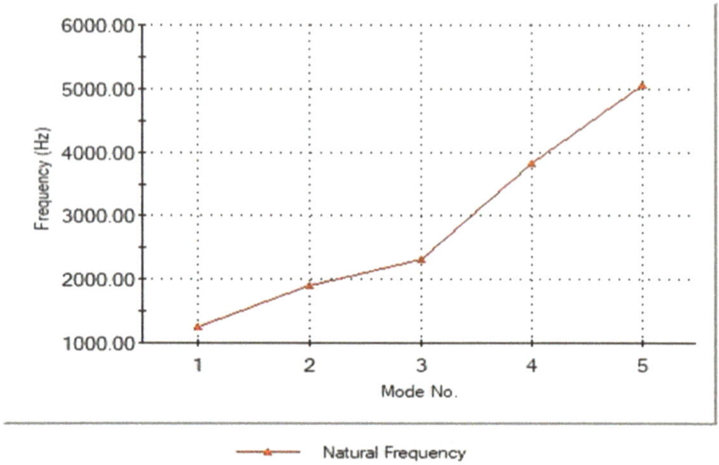

Fig. 8.24 Frequency response of the optimized link 2 subassembly 1. *(Data from Marvellous Moyo and Tawanda Mushiri.)*

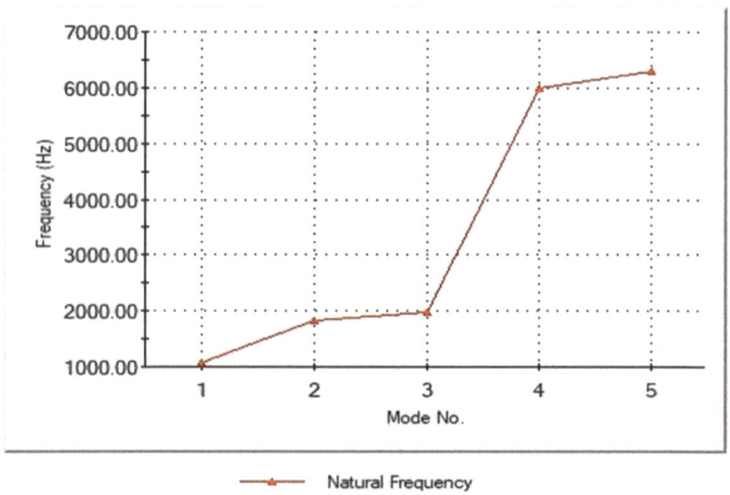

Fig. 8.25 Frequency response of the optimized link 3. *(Data from Marvellous Moyo and Tawanda Mushiri.)*

8.5 Discussion

The total original total mass of the manipulator components before assembly was 2.43 kg and after optimization it was 2 kg. Therefore the results show that there was a mass reduction of 17.76% for the optimized robotic manipulator. Reducing the weight of a robotic manipulator improves the joint

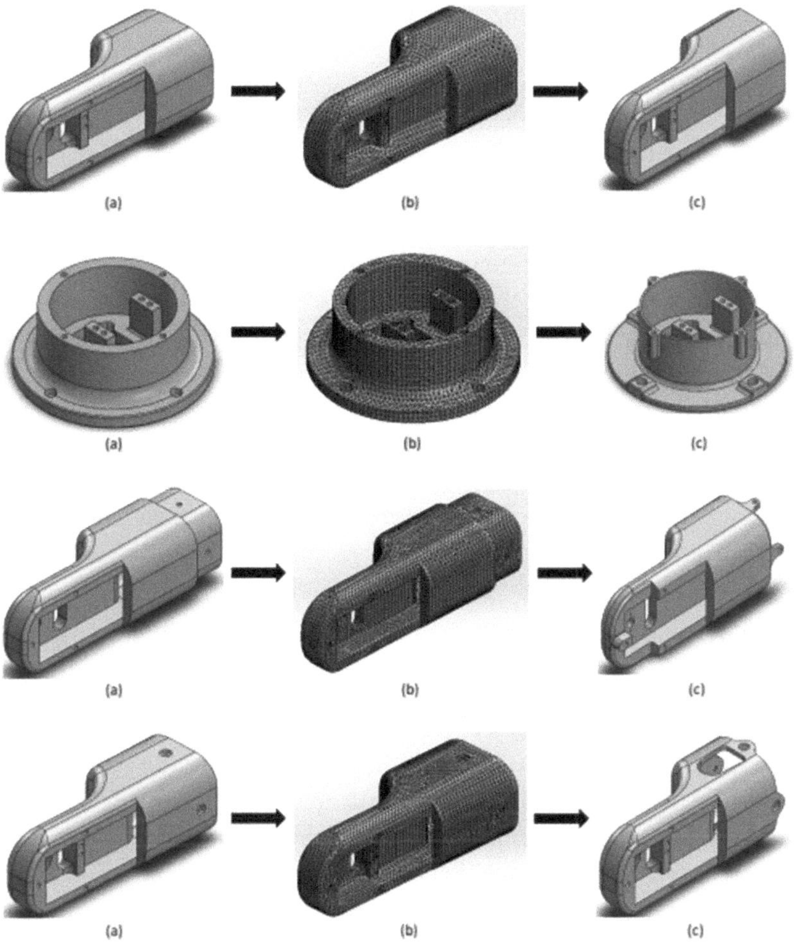

Fig. 8.26 Topology optimization of the robotic manipulator original components (A) through topology meshed result under static loading (B) to the optimized model (C). *(Data from Marvellous Moyo and Tawanda Mushiri.)*

torque efficiencies and hence the overall performance of the whole system. Medical robotic manipulators are usually very expensive owing to the technology that is involved and therefore, one way to reduce costs can be material reduction of the robotic manipulator's components.

The maximum stresses for the original components are slightly less than those experienced by the optimized components and all of these stresses are way below the yield strength of Kevlar 53%. Therefore the components will not elastically fail under the applied loads specified in Table 8.1. In addition, factor of safety of the optimized components though it is below that of the original component still it is of high magnitude. This shows that the robotic manipulator components have been overdesigned, which is a good attribute for medical robotic manipulator, as failure might result in loss of human life.

The total displacement for the original base link was of 2.241×10^{-4} mm which is less than that of optimized link which was 2.242×10^{-4} mm. In can be seen that for all of the links under consideration, there is a slight variation between the total displacements of the original and optimized base link. Therefore this implies that the optimized base link still has sound structural integrity even after 17.76% mass reduction.

Modal analysis results show that the first fundamental frequencies for all of the optimized robotic manipulator component designs were higher than the forcing frequency due to the servo motors. Therefore for these types of servo motor, resonance will not transpire. In addition, these natural frequencies can be used to avoid resonance during testing since many uncertainties can occur.

8.6 Conclusion

Medical robots play a very significant role in medicine, especially during a pandemic such as COVID-19 virus, but they must be very safe, reliable, and economical. One way to address all of these factors is through design optimization of the robotic manipulator's structural components. In this study, an attempt has been performed to optimize the robotic manipulator components' structure using finite element analysis. Static and modal analyses were done in order to contrast the stress and displacement distributions between the original components and optimized components using topology. An optimal lightweight component was obtained through the use of topology optimization. The static and modal analyses confirmed the robotic manipulator components' degree of safety. Study findings contribute to the ongoing design of an intelligent machine for testing COVID-19 virus, by providing information that will be used for components selection and life cycle costing of the whole system.

References

Alric, M., et al. (2009). Potential applications of medical and non-medical robots for neurosurgical applications. In *Minimally invasive therapy and allied technologies* (pp. 193–216). https://doi.org/10.1080/13645700903053584.

Bendel, O. (2015). Surgical, therapeutic, nursing and sex robots in machine and information ethics. In *Vol. 74. Intelligent systems, control and automation: Science and engineering* (pp. 17–32). Kluwer Academic Publishers. https://doi.org/10.1007/978-3-319-08108-3_2.

Challacombe, B. J., et al. (2006). The history of robotics in urology. *World Journal of Urology, 24*(2), 120–127. https://doi.org/10.1007/s00345-006-0067-1.

Ewing, D. R., et al. (2004). Robots in the operating room—The history. *Seminars in Laparoscopic Surgery, 11*(2), 63–71. https://doi.org/10.1177/107155170401100202. Westminster Publications Inc.

Fatima, S., et al. (2021). Exploring the significant applications of Internet of Things (IoT) with 3D printing using advanced materials in medical field. *Materials Today: Proceedings*. Elsevier. Available at: https://www.sciencedirect.com/science/article/pii/S221478532 1003941 (Accessed 13 May 2021).

Girone, M., et al. (2001). Stewart platform-based system for ankle telerehabilitation. *Autonomous Robots, 10*(2), 203–212. https://doi.org/10.1023/A:1008938121020. Kluwer Academic Publishers.

Goher, K. M., Mansouri, N., & Fadlallah, S. O. (2017). Assessment of personal care and medical robots from older adults' perspective. *Robotics and Biomimetics, 4*(1). https://doi.org/10.1186/s40638-017-0061-7. Springer Science and Business Media LLC.

Hagn, U., et al. (2008). The DLR MIRO: A versatile lightweight robot for surgical applications. *Industrial Robot, 35*(4), 324–336. https://doi.org/10.1108/01439910810876427.

Hilder, J., et al. (2016). The PSI swarm: A low-cost robotics platform and its use in an education setting. In *Lecture notes in computer science (including subseries lecture notes in artificial intelligence and lecture notes in bioinformatics)* (pp. 158–164). Springer Verlag. https://doi.org/10.1007/978-3-319-40379-3_16.

JSUMO. (2021). *MG996R servo motor (digital) servo motors TowerPro*. JSumo.com. Available at: https://www.jsumo.com/mg996r-servo-motor-digital (Accessed 13 May 2021).

Kalo, A., & Newsum, M. J. (2014). An investigation of robotic incremental sheet metal forming as a method for prototyping parametric architectural skins. In *Robotic fabrication in architecture, art and design 2014* (pp. 33–49). Springer International Publishing. https://doi.org/10.1007/978-3-319-04663-1_3.

Khatami, S. (2002). *Isotropic design optimization of robotic manipulators using a genetic algorithm method*. ieeexplore.ieee.org. Available at: https://ieeexplore.ieee.org/abstract/document/1157824/ (Accessed 13 May 2021).

Mavroidis, C. (2002). Development of advanced actuators using shape memory alloys and electrorheological fluids. *Research in Nondestructive Evaluation, 14*(1), 1–32. https://doi.org/10.1080/09349840209409701.

Nawrat, Z. (2006). *DZIELIMY SIĘ DOŚWIADCZENIEM Medical robots in cardiac surgery*. Kardiochirurgia i Torakochirurgia Polska/Polish. Available at: https://www.termedia.pl/DZIELIMY-SIE-DOSWIADCZENIEM-Medical-robots-in-cardiac-surgery,40,11585,0,1.html (Accessed 13 May 2021).

Oliwa, R. (2020). The mechanical properties of kevlar fabric/epoxy composites containing aluminosilicates modified with quaternary ammonium and phosphonium salts. *Materials, 13*(17). https://doi.org/10.3390/MA13173726. MDPI AG.

Polygerinos, P., et al. (2017). Soft robotics: Review of fluid-driven intrinsically soft devices; manufacturing, sensing, control, and applications in human-robot interaction. *Advanced Engineering Materials*. https://doi.org/10.1002/adem.201700016. Wiley-VCH Verlag.

Roy, A., Ghosh, T., & Mishra, R. (2018). *Dynamic FEA analysis and optimization of a robotic arm for CT image guided procedures*. Elsevier. Available at: https://www.sciencedirect.com/science/article/pii/S2214785318314081 (Accessed 13 May 2021).

Rupal, B. S., et al. (2017). Lower-limb exoskeletons: Research trends and regulatory guidelines in medical and non-medical applications. *International Journal of Advanced Robotic Systems*. https://doi.org/10.1177/1729881417743554. SAGE Publications Inc.
Scott, K. (2020). *The investigation of low cost computer vision application for first responder co-robotics*. Available at: https://repository.asu.edu/items/62627 (Accessed 13 May 2021).
Srinivas, G. (2020). *Topology optimization of rigid-links for industrial manipulator considering dynamic loading conditions*. Elsevier. Available at: https://www.sciencedirect.com/science/article/pii/S0094114X20302007 (Accessed 13 May 2021).
Tang, Y., Yu, Y., & Shi, J. (2017). *Modal and harmonic response analysis of key components of robotic arm based on ANSYS*. jvejournals.com. Available at: https://www.jvejournals.com/article/18703 (Accessed 13 May 2021).
Taylor, R. H., et al. (2016). Medical robotics and computer-integrated surgery. In *Springer handbook of robotics* (pp. 1657–1683). Springer International Publishing. https://doi.org/10.1007/978-3-319-32552-1_63.
van Wynsberghe, A. (2013). Designing robots for care: Care centered value-sensitive design. *Science and Engineering Ethics*, *19*(2), 407–433. https://doi.org/10.1007/s11948-011-9343-6.
Wolf, A., & Shoham, M. (2009). Medical automation and robotics. In *Springer handbook of automation* (pp. 1397–1407). Berlin Heidelberg: Springer. https://doi.org/10.1007/978-3-540-78831-7_78.
Yap, H. K., Chen, R., & Yeow, H. (2016). High-force soft printable pneumatics for soft robotic applications soft wearable robotics for rehabilitation and assistive applications view project wellness, ageing, serious games and healthcare communities view project. *Soft Robotics*, *3*(3), 144–158. https://doi.org/10.1089/soro.2016.0030. liebertpub.com. Mary Ann Liebert Inc.
Zhang, Y., & Lu, M. (2020). A review of recent advancements in soft and flexible robots for medical applications. *International Journal of Medical Robotics and Computer Assisted Surgery*. https://doi.org/10.1002/rcs.2096. John Wiley and Sons Ltd.
Zhou, L., et al. (2012). Integrated dimensional and drive-train design optimization of a light-weight anthropomorphic arm EXOTIC: Exoskeleton arm with tongue based intelligent control view project integrated dimensional and drive-train design optimization of a light-weight anthropomorphic arm. *Robotics and Autonomous Systems*, *60*, 113–122. https://doi.org/10.1016/j.robot.2011.09.004.

Further reading

Kindel Media. (2021). Full shot of robot toy. In *Studies in Health Technology and Informatics*. https://doi.org/10.3233/978-1-60750-706-2-716.

CHAPTER 9

Life cycle analysis of medical robotic manipulators

9.1 Introduction

Sustainability and environmental effect have rapidly become crucial issues in the manufacturing business and the consciousness of green manufacturing has driven engineers to be increasingly more environmentally cognizant (Hsu et al., 2013; Lee et al., 2005; Millar & Russell, 2011). Owing to their accuracy and precision, the robotic tools have been generally utilized in different manufacturing enterprises (Hägele et al., 2016; Pan et al., 2012). In spite of the fact that they are by and large treated as supporters of the green manufacturing since use of these robotic tools has fundamentally improved the productivity and decreased the energy and material utilization, the specific part of the robotic instruments in the green manufacturing actually should be altogether examined (Garetti & Taisch, 2012; Gude et al., 2013). In this chapter, a life cycle assessment of various concepts of a medical robotic manipulator that will be used for testing COVID-19 virus will be conducted. The environmental aspects and potential effects related with the premanufacturing, manufacturing, use, and end of life of the robotic instruments for testing COVID-19 virus will be evaluated.

One of the quickest developing field of robotics in medical care concerns surgeons and their profession at the highest point of the clinical natural way of life (Compagni, Mele, & Ravasi, 2015; Mukherjee & Tripathy, 2021). Surgical robots are the wonders of a medical procedure. As indicated by market investigation, the business is going to blast (Evans, Medina, & Dwyer, 2018; Haidegger et al., 2020; Mende et al., 2019; Staiger, 2009). While it is hard to check the takeup of robotics inside the clinical manufacturing industry, figures do show that robot sales to the electronics business, which incorporates clinical gadgets, have risen essentially since 2013 (Keeley et al., 2013; Maghazil & Yellowlees, 2014; Sathya et al., 2021).

This is because robotics offers manufacturers an approach to smooth out production lines, assisting with increasing productivity and streamline work processes (Azizi & Yazdi, 2018; Madakam & Holmukhe, 2019). Inside

https://doi.org/10.1016/B978-0-323-99443-9.00001-2

clinical manufacturing, robots are being utilized across the whole production line—from assembly to investigation and packaging. Owing to the severe guidelines, device manufacturers are feeling the squeeze to reproduce a product, under similar tough conditions, with little downtime (Bruce & Olshfski, 2019; Burawoy, 2012; Chandrasekaran, 2009). To stay aware of the speed of advancement, clinical manufacturers should have the option to change fabricating lines when new items are waiting to be created (Dove, 2002; Gomez-Marquez & Hamad-Schifferli, 2019). Robotics and automation offer makers modular frameworks that advance productivity and work process. Robots have been utilized in the medical care sector for quite a while, working generally in the background (Mokyr, Vickers, & Ziebarth, 2015; Stahl & Coeckelbergh, 2016). In the course of the most recent 5 years, the scope of robotic applications in medical field has extended quickly to incorporate assistive applications for physicians, caregivers, nurses, and patients in emergency clinics and care facilities.

9.2 Literature review

An investigation on life cycle analysis of dual-arm type robotic apparatus for LCD panel handling was done by Wyatt et al. (2017). The investigation incorporated a daily existence cycle evaluation of dual-arm type robotic apparatus made with three unique materials. The mechanical arms were utilized for LCD panel handling. Environmental effects produced during all the life cycle of the robots were analyzed. The examination showed that composite materials have minimum environmental effect contrasted with metallic materials. It was likewise tracked down that the main effect category created by the robotic apparatus is cancer-causing agent, while the utilization phase of the robotic instrument's life cycle has the best environmental effect.

According to Yeniaras, Deng, Syed, Davies, and Tsekos (2011), advancement of image-guided and robot-assisted systems can be helpful to intracardiac mediations. Their paper proposed a novel methodology and an augmented simulation framework for preoperative arranging and intraoperative direction of cardiovascular systems, and for examining the kinematics and control of virtual robotic manipulators, in light of MRI CINE images. The framework joins dedicated programming modules for processing MR images, creating dynamic trajectories in the ceaselessly changing environment of a beating heart, controlling a particular nonexclusive virtual manipulator along those directions, and an augmented reality interface that melds every one of those data. The proposed framework was applied for the recreation of getting to the aortic valve annulus by means of a little entry point on the apex by moving a robotic manipulator through an entrance passageway that securely crosses over the left ventricle (LV) of the beating heart.

According to Zhao, Li, and Gao (2008, 2009), another terminal sliding mode control approach is created for robotic manipulators dependent on finite-time stability hypothesis and differential inequality principle. The comparing stability investigation was introduced to establish a framework for hypothetical comprehension to the hidden design issue just as safe activity for genuine framework. An illustrative illustration of a two-interface rigid robotic manipulator was introduced to approve adequacy of the proposed approach.

Chung, Wang, and Cooper (2013) led a literature search of assistive robotic manipulators from 1970 to 2012 in PubMed, Google Scholar, and University of Pittsburgh Library Framework—PITTCat. Twenty pertinent investigations were distinguished. Studies were isolated into two general classes: user task inclinations and UI execution estimations of marketed and creating assistive robotic arms. The result measures and ICF codes related with the presentation assessments were reported. Ideas for the future examinations incorporated: normalized ADL undertakings for the quantitative and qualitative assessment of task proficiency and execution to assemble comparable measures between research gatherings, contemplates relevant to the assignments from user need records and ICF codes, and fitting clinical practical evaluation tests with thought of imperatives in assistive robotic arm user interfaces. What's more, these result estimates will help doctors and specialists fabricate normalized apparatuses while recommending and surveying assistive robotic manipulators.

9.3 Materials and methods

9.3.1 Aim and scope

The principal target of this investigation is to give a Life Cycle Analysis (LCA) for the robotic manipulator which can collect and test COVID-19 virus samples without human intervention. To completely depict the material and energy necessities for each robotic manipulator alternative, this investigation utilized cited values or determined information to lead an examination.

Functional unit is set as years of utilization, which is a standard unit of measure when administration conveyed is estimated by time. In this examination, the normal service life for the automated arms is thought to be 8 years, which depends on the normal service life of industrial robots (Engleberger, 2012). The objectives of this LCA case study include: to comprehend which eco-design techniques bring about the best environmental execution improvement, to distinguish which stage will represent the best environmental effect during the entire existence of the robotic manipulator, and to decide how the outcomes can be utilized to more readily characterize the functional prerequisites and design parameters for the robotic manipulator in the product acknowledgment process.

9.3.2 Framework boundaries

In this examination, the robotic manipulator used to collect COVID-19 samples is assumed to be manufactured in Asia, since most of the required resources are available in this continent. The fabricated robotic manipulators are then moved by truck to airport in Asia and transported to Zimbabwe by air; at long last, they will be conveyed by truck to Harare Central Hospital. Note that the place of manufacture, delivery, and mode of transport have been assumed for the purposes of study completion, since the manipulators are still in design phase. The medical robotic manipulators are required to stay in the Harare Central Hospital in Zimbabwe for their service life, assessed to be 8 years. Table 9.1 presents the distance from the location of the manufacture of the robotic manipulators to the area of its user, the medical healthcare facility. Also, note due to the sustainability constraint of SolidWorks, it has been assumed that the medical robotic manipulator, for the purpose of study completion, will be used in North America during its life cycle.

Table 9.1 Distances from the manufacturer of the medical robot to the hospital.

Description	Mode of transport	Distance
From the manufacture to the airport in the United States	Truck	10 km
From US airport to RG Mugabe International Airport	Cargo aircraft	14,321 km
From RG Mugabe International Airport to Harare Central Hospital	Truck	16.5 km

This LCA study will assess the environmental effects related with all life phases of the robotic manipulator, which incorporate the premanufacturing stage, manufacturing of the robotic manipulator, use, just as disposal, reuse or recycle of the robotic manipulator. To work on the examination, the mechanical segments of the robot will be included for the investigation, while electrical segments (wiring, sensors) will be barred because of their generally little sizes and similitude. Power utilizations of the robotic manipulators will be determined dependent on the expected prerequisite of a fittingly estimated sized motor which can satisfy the functional necessities of robotic manipulator.

9.3.3 Materials

Robotic manipulators developed with two different materials are investigated in this research. Table 9.2 records the material breakdown of the significant parts of a robot which is created with composite materials, namely Kevlar 53%.

Table 9.3 presents the same components but now carbon fiber is used as the material.

Table 9.2 Material breakdown of the medical robotic manipulator using Kevlar 53%.

Component	Mass (kg)	Materials
Base link	1.31	Kevlar 53%
Link1	1.23	Kevlar 53%
Link 2	0.61	Kevlar 53%
Link 3	0.46	Kevlar 53%
Link 4	0.34	Kevlar 53%
End effector	0.24	Kevlar 53%

Table 9.3 Material breakdown of the medical robotic manipulator using carbon fiber.

Component	Mass (kg)	Materials
Base link	1.55	Carbon fiber
Link1	1.46	Carbon fiber
Link 2	0.72	Carbon fiber
Link 3	0.55	Carbon fiber
Link 4	0.41	Carbon fiber
End effector	0.28	Carbon fiber

9.3.4 Manufacturing and bill of materials

The manufacturing cycles of the significant segments, such as end effectors, wrists, links, and the base, have been summarized in this section. The bill of materials has been divided into custom part costing and the other readily available quantities. Following is a report showing an estimation of the costs which may be incurred without necessarily comparing with the said benefits. The bill of quantities is summarized in the following table and all prices were estimated from Alibaba. Note should be taken that only the cost of the robotic manipulator (under custom costing section) has been done, while rest of the components such as the booth components have been excluded. Table 9.4 also includes the type of manufacturing process for each part.

Table 9.4 Bill of materials.

Component	Component material	Manufacturing method	Quantity	Part cost (USD/part)	Total cost (USD)
Custom parts					
Base Cover	Kevlar epoxy 53%	3D Printing	1	5	5
Base	Kevlar epoxy 53%	3D Printing	1	7	7
Big servo cover	Kevlar epoxy 53%	3D Printing	4	5	20
End effector back cover	Kevlar epoxy 53%	3D Printing	1	5	5
End effector top cover	Kevlar epoxy 53%	3D Printing	1	5	5
Link 1 subassembly 1 cover	Kevlar epoxy 53%	3D Printing	1	5	5
Link 1 subassembly 1	Kevlar epoxy 53%	3D Printing	1	7	7
Link 1 subassembly 2	Kevlar epoxy 53%	3D Printing	1	7	7
Link 2 subassembly 1	Kevlar epoxy 53%	3D Printing	1	7	7
Link 3 cover	Kevlar epoxy 53%	3D Printing	1	5	5
Link 3	Kevlar epoxy 53%	3D Printing	1	7	7
Small servo cover	Kevlar epoxy 53%	3D Printing	2	5	10

Table 9.4 Bill of materials—cont'd

Component	Component material	Manufacturing method	Quantity	Part cost (USD/part)	Total cost (USD)
Spherical wrist link 1 cover	Kevlar epoxy 53%	3D Printing	1	5	5
Spherical wrist link 1	Kevlar epoxy 53%	3D Printing	1	7	7
Spherical wrist link 2 cover	Kevlar epoxy 53%	3D Printing	1	5	5
Spherical wrist link 2	Kevlar epoxy 53%	3D Printing	1	7	7
Other					
MG996R servo	–	–	4	$30/4 servo	30
SG90 Micro servo	–	–	3	$15/3 servo	15
Raspberry pi 3 model B+	–	–	1		35
1.5 mm 3 core flex cable	–	–	5 m	$1/m	5
Bolts and nuts	Stainless steel	–	Lump sum	10	10
TOTAL					**209**

9.3.5 Power utilization

The power consumption of robotic manipulators used to test COVID-19 virus is determined dependent on the rated power of the drive motors of the arms and wrists. Since both materials used are lightweight composites, then same motors will be used for both the robotic manipulators. The life activity hours depend on the understanding of 22 h of activity each day, for an absolute 8-year life range. Table 9.5 records the determined life energy utilizations of the motors of these robots.

9.3.6 End of life

Composite materials represent a challenge for recycling, as they cannot be economically isolated into their segment classifications. In this investigation, most of the composite materials (carbon fiber) utilized for the development of the robotic component (links, wrists, and end effector) are thought to be shipped off to a landfill toward the finish of the robot's service life while a

Table 9.5 Power consumptions of the medical robots.

Part	Selected motor	Rated power (W)	Material type	Life operation	Total energy use (kWh)
Joint 1	MG996R servo	12.18	Composite	64,240	78.24
Joint 2	MG996R servo	393.69	Composite	64,240	2529.06
Joint 3	MG996R servo	591.54	Composite	64,240	3800.05
Joint 4	MG996R servo	10.38	Composite	64,240	66.68
Joint 5	SG90 Micro servo	98.54	Composite	64,240	633.02
Joint 6	SG90 Micro servo	11.11	Composite	64,240	71.37

small percentage is recycled. The robotic manipulator made using Kevlar 53% can be recycled since this composite is 100% recyclable.

9.4 Results

Considering the earlier assumptions and formulations, a LCA model was designed using SolidWorks 2018 sustainability tools, and the environmental impacts produced by the robotic manipulator can be calculated. Fig. 9.1 shows the overall impact of the robotic manipulator made with Kevlar 53% while Fig. 9.2 shows the same robotic manipulator now made using carbon fiber.

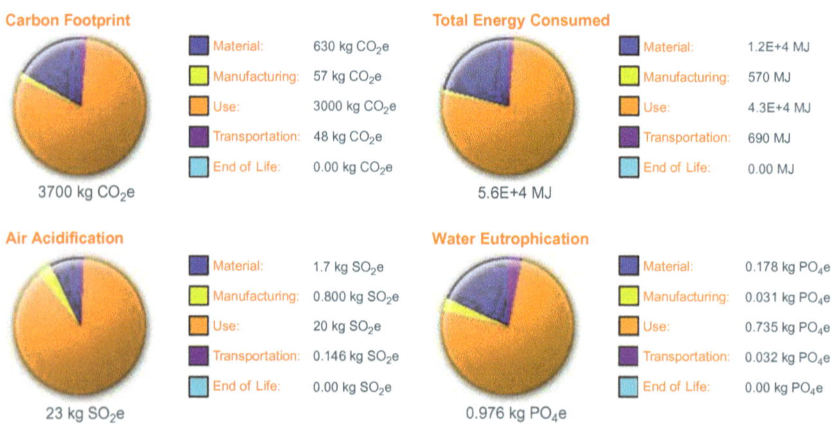

Fig. 9.1 Total impact of the robotic manipulator made with Kevlar 53%. *(Courtesy Marvellous Moyo and Tawanda Mushiri.)*

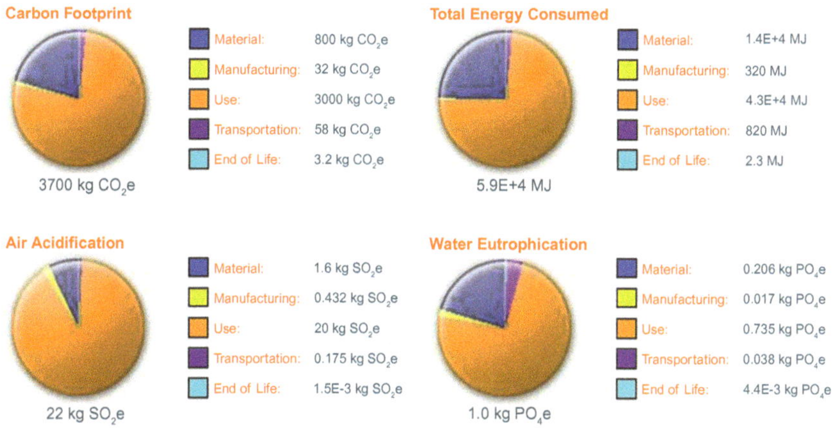

Fig. 9.2 Total impact of the robotic manipulator made with carbon fiber. *(Courtesy Marvellous Moyo and Tawanda Mushiri.)*

The environmental impacts have been classified into four main categories, namely, carbon foot print, total energy consumed, air acidification, and water eutrophication. Carbon foot print has the largest impact on the environment as shown in the earlier figures. Each category is divided into five subcategories, namely, material, manufacturing, use, transportation, and end of life. For both the materials, the largest impact subcategory is energy used during the lifetime of the robotic manipulator, followed by the material and transport mode used. Manufacturing and end of life contribute less to the overall environmental impact of the robotic manipulator.

Utilizing 8 years as the operational unit (the anticipated useful life of the robotic manipulator), the total mass of equivalent carbon dioxide emitted by the manipulator during its life was evaluated. The total carbon foot print is 37 kg when Kevlar 53% was used as the material of the robotic manipulator and 39 kg when carbon fiber was used as the material. The energy consumed for both materials was assumed to be constant while the manipulator made with Kevlar 53% had excess of 1 kg compared to the one made with carbon fiber in terms of air acidification. Kevlar 53% manipulator had 0.976 kg of water eutrophication while the carbon fiber had 1 kg.

9.5 Discussion

Environmental impact comparison of the two materials has shown that the greatest impact category is carbon foot print which has the highest effect on the climate change, environment, and human health. This sustainability study shows that recyclable material like Kevlar 53% generates less carbon foot print compared to those that are not. Therefore it is advisable to employ such material during design of medical robotic manipulators, as this will help save human lives in future and preserve the environment.

The results show that the material choice of the medical robotic manipulator has the greatest environmental impact with significant consequences, which will be difficult to determine if the life cycle analysis of the robot was not done. This is because the type of material affects energy used during the life of product, transportation, and manufacturability, which in turn will indirectly affect the environment through the increase of carbon foot print, air acidification, and water eutrophication. From the earlier discussion it can be seen that life cycle analysis is necessary for the design process of the medical robotic manipulator, as design has several environmental impacts during its life cycle.

9.6 Conclusions

LCA analysis of the medical robotic manipulator developed with two different composite materials, such as Kevlar 53% and carbon fiber, has revealed that the most vital impact produced by the medical robotic arm is carbon foot print. Also, it was shown that the use stage of the medical robotic manipulator has the highest contribution to this high value of carbon foot print and energy consumed leading to even greater environmental pollution and global warming. Medical robotic manipulators are designed to help people in healthcare facility setups and in this case the robot will be specifically used for testing COVID-19 virus in hospitals, so it is crucial to know how they will affect the same lives of people during their operational life. Thus it was important to perform the LCA and it was inferred that medical robotic manipulator with higher efficiency, lighter weight, and recyclable material will produce lower environmental impact. This study findings will contribute to the ongoing study to design an intelligent robotic manipulator used to test COVID-19 virus in the material selection process.

References

Azizi, A., & Yazdi, P. G. (2018). Design and fabrication of intelligent material handling system in modern manufacturing with industry 4.0 approaches. *International Robotics & Automation Journal*, *4*(3). https://doi.org/10.15406/iratj.2018.04.00121. Researchgate.net. MedCrave Group, LLC.

Bruce, W., & Olshfski, D. (2019). The new American workplace. In *Public productivity handbook* (2nd ed., pp. 488–505). (2nd ed.). Taylor and Francis. https://doi.org/10.7591/9781501720642 Revised and Expanded.

Burawoy, M. (2012). *Manufacturing consent.* Available at:https://www.degruyter.com/document/doi/10.7208/9780226217710/html (Accessed 23 June 2021).
Chandrasekaran, C. (2009). *Rubber seals for fluid and hydraulic systems.* Available at:https://books.google.com/books?hl=en&lr=&id=COgvEhUnFq8C&oi=fnd&pg=PP1&dq=*Owing+to+the+severe+guidelines,+device+manufacturers+are+feeling+the+squeeze+to+reproduce+a+product,+under+similar+tough+conditions,+with+little+downtime.+&ots=AmnWzLkpMf&sig=HHPYeW3nuaSzrf5lpbYl3gGUOdQ (Accessed 23 June 2021).
Chung, C. S., Wang, H., & Cooper, R. A. (2013). Functional assessment and performance evaluation for assistive robotic manipulators: Literature review. *Journal of Spinal Cord Medicine*, 273–289. https://doi.org/10.1179/2045772313Y.0000000132.
Compagni, A., Mele, V., & Ravasi, D. (2015). How early implementations influence later adoptions of innovation: Social positioning and skill reproduction in the diffusion of robotic surgery. *Academy of Management Journal*, *58*(1), 242–278. https://doi.org/10.5465/amj.2011.1184.
Dove, R. (2002). *Response ability: The language, structure, and culture of the agile enterprise.* Available at: https://books.google.com/books?hl=en&lr=&id=M3wnwrol2YgC&oi=fnd&pg=PR17&dq=*To+stay+aware+of+the+speed+of+advancement,+clinical+manufacturers+should+have+the+option+to+change+fabricating+lines+when+new+items+are+waiting+to+be+created.+Robotics+and+automation+offer+makers+modular+frame works+that+advance+productivity+and+work+proce&ots=e_wWEEWCqx&sig=ZROnFdf3nUP4txk3PA7ndzzwXk4 (Accessed 23 June 2021).
Engleberger, J. F. (2012). *Robotics in practice: Management and applications of industrial robots. Joseph F. Engelberger—Google Books.* Available at:https://books.google.co.zw/books?hl=en&lr=&id=r1Z-BgAAQBAJ&oi=fnd&pg=PT9&dq=In+this+examination,+the+normal+service+life+for+the+automated+arms+is+thought+to+be+eight+years,+which+depends+on+the+normal+service+life+of+industrial+robots+%5B6%5D&ots=9w68BKxrLy&sig=c_DUf51S_4SdpA__o7s5Iu8d5J4&redir_esc=y#v=onepage&q&f=false (Accessed 23 June 2021).
Evans, C. R., Medina, M. G., & Dwyer, A. M. (2018). Telemedicine and telerobotics: From science fiction to reality. In *Updates in surgery* (pp. 357–362). Springer-Verlag Italia Srl. https://doi.org/10.1007/s13304-018-0574-9.
Garetti, M., & Taisch, M. (2012). Sustainable manufacturing: Trends and research challenges. *Production Planning and Control*, *23*(2–3), 83–104. https://doi.org/10.1080/09537287.2011.591619.
Gomez-Marquez, J., & Hamad-Schifferli, K. (2019). Distributed biological foundries for global health. In *Advanced Healthcare Materials.* Wiley-VCH Verlaghttps://doi.org/10.1002/adhm.201900184.
Gude, V., et al. (2013). Microwave energy potential for biodiesel production. *Sustainable Chemical Processes*, *1*(1), 5. Springer Naturehttps://doi.org/10.1186/2043-7129-1-5.
Hägele, M., et al. (2016). Industrial robotics. In *Springer handbook of robotics* (pp. 1385–1421). Springer international publishing. https://doi.org/10.1007/978-3-319-32552-1_54.
Haidegger, T., et al. (2020). *Industrial and medical cyber-physical systems: Tackling user requirements and challenges in robotics*: (pp. 253–277). Springer. https://doi.org/10.1007/978-3-030-14350-3_13.
Hsu, C.-C., et al. (2013). Supply chain drivers that foster the development of green initiatives in an emerging economy. *International Journal of Operations & Production Management*, *33*(6), 656–688. https://doi.org/10.1108/IJOPM-10-2011-0401 emerald.com.
Keeley, L., et al. (2013). *Ten types of innovation: The discipline of building breakthroughs.* Available at:https://books.google.com/books?hl=en&lr=&id=TW4YAAAAQBAJ&oi=fnd&pg=PT14&dq=*While+it+is+hard+to+check+the+take-up+of+robotics+inside+the+clinical+manufacturing+industry,+figures+do+show+that+robot+sales+to+the+electronics+business,+which+incorporates+clinical+gadgets,+have+risen

+essentially+since+2013.+&ots=oU4OVJn9kd&sig=8sYu0nEX7l4ep8TD30_nN9l6aZY (Accessed 23 June 2021).
Lee, S.-G., et al. (2005). Design for the environment: Life cycle assessment and sustainable packaging issues. Smart manufacturing automation View project Cyber-Physical Machine Tools for Smart Manufacturing View project*International Journal of Environmental Technology and Management*, *5*(1), 14–41. https://doi.org/10.1504/IJETM.2005.006505 Inderscience Publishers.
Madakam, S., & Holmukhe, R. (2019). *The future digital work force: Robotic process automation (RPA)*. SciELO Brasil. Available at:https://www.scielo.br/scielo.php?pid=S1807-17752019000100300&script=sci_arttext (Accessed 23 June 2021).
Maghazil, A., & Yellowlees, P. (2014). Novel approaches to clinical care in mental health: From asynchronous telepsychiatry to virtual reality. *Studies in Computational Intelligence*, *491*, 57–78. https://doi.org/10.1007/978-3-642-38550-6_4 Springer Verlag.
Mende, M., et al. (2019). Service robots rising: How humanoid robots influence service experiences and elicit compensatory consumer responses. *Journal of Marketing Research*, *56*(4), 535–556. https://doi.org/10.1177/0022243718822827 SAGE Publications Ltd.
Millar, H. H., & Russell, S. N. (2011). The adoption of sustainable manufacturing practices in the Caribbean. *Business Strategy and the Environment*, *20*(8), 512–526. https://doi.org/10.1002/bse.707.
Mokyr, J., Vickers, C., & Ziebarth, N. L. (2015). The history of technological anxiety and the future of economic growth: Is this time different? *Journal of Economic Perspectives*, 31–50. https://doi.org/10.1257/jep.29.3.31 American Economic Association.
Mukherjee, D., & Tripathy, H. (2021). *Scope of medical bots in clinical domain*. Springer. Available at:https://link.springer.com/content/pdf/10.1007/978-981-33-4698-7.pdf#page=345 (Accessed 23 June 2021).
Pan, Z., et al. (2012). *Recent progress on programming methods for industrial robots*. Recommended CitationElsevier. Available at: https://ro.uow.edu.au/eispapershttps://ro.uow.edu.au/eispapers/29 (Accessed 23 June 2021).
Sathya, V., et al. (2021). An obfuscation technique for malware detection and protection in sandboxing'. In *Artificial intelligence for cyber security: Methods, issues and possible horizons or opportunities* (pp. 235–261). . https://doi.org/10.1007/978-3-030-72236-4_10.
Stahl, B. C., & Coeckelbergh, M. (2016). Ethics of healthcare robotics: Towards responsible research and innovation. *Robotics and Autonomous Systems*, *86*, 152–161. https://doi.org/10.1016/j.robot.2016.08.018 Elsevier B.V.
Staiger, D. (2009). The recent surge in nurse employment: Causes and implications: Recession effects that have eased the shortage of hospital nurses must be viewed as temporary, lest they distract policymakers from continuing to address longer-term indicators. *Health Affairs*. *28*(4)https://doi.org/10.1377/hlthaff.28.4.w657 healthaffairs.org.
Wyatt, H., et al. (2017). Life cycle analysis of double-arm type robotic tools for LCD panel handling. *Machines*, mdpi.com. Available at:https://www.mdpi.com/2075-1702/5/1/8 (Accessed 23 June 2021).
Yeniaras, E., Deng, Z., Syed, M. A., Davies, M. G., & Tsekos, N. V. (2011). A novel virtual reality environment for preoperative planning and simulation of image guided intracardiac surgeries with robotic manipulators. *Studies in health technology and informatics*, *163*, 716–722. https://doi.org/10.3233/978-1-60750-706-2-716 [Ebook Volume 163: Medicine Meets Virtual Reality 18].
Zhao, D., Li, S., & Gao, F. (2008). *A new terminal sliding mode control for robotic manipulators system dynamic model*. Elsevierhttps://doi.org/10.3182/20080706-5-KR-1001.1373.
Zhao, D., Li, S., & Gao, F. (2009). A new terminal sliding mode control for robotic manipulators. *International Journal of Control*, *82*(10), 1804–1813. https://doi.org/10.1080/00207170902769928.

CHAPTER 10

Future of intelligent robots for testing specimens

10.1 Introduction

According to WHO, COVID-19 pandemic is at present a worldwide health emergency: in excess of 179 million cases had been affirmed worldwide and in excess of 3.87 million deaths had happened at the time of composing this report (WHO, 2021). Quick and precise recognition of COVID-19 is progressively crucial to forestall the sources of infections just as assisting patients with deterring disease progression. Soft computing strategies, such as fuzzy logic, neural networks, and genetic algorithms, have demonstrated as possible tools in the disease identification. It is against this background that the researchers proposed the design of an intelligent robotic machine for testing COVID-19 virus to be installed in hospitals.

This project report has been broken down into 10 chapters. First chapter is an introduction on the use of artificial intelligence and robotics in healthcare, including the project problem statement, aim, objectives, and justification. After that, an in-depth literature survey on the use of robotics and artificial intelligence in healthcare was done. Chapter 3 covered the robotic components' description and main materials and methods used to solved the stated problem and achieve the aims and objectives. A study to determine COVID-19 effects and current testing techniques were done in Chapters 4 and 5 covered the development considerations of an intelligent booth that will be used during COVID-19 testing. Chapter 6 covered the use of robotics, artificial intelligence, and deep learning to collect COVID-19 samples while Chapter 7 covered the validation of the robotic manipulator kinematics and dynamics using the MATLAB robotics toolbox. In Chapter 8, the design optimization of the robotic manipulator was done. Lastly, the life cycle analysis was explored in Chapter 9 and this chapter is the summary and recommendations of the research. At the end, there are Appendices A–C.

Healthcare Systems Design of Intelligent Testing Centers
https://doi.org/10.1016/B978-0-323-99443-9.00004-8

10.2 Limitations and the way forward

10.2.1 Prototyping and testing

The aims and objectives of this book, as stated in Chapter 1, were all achieved except prototyping and testing. Prototype testing is a very crucial part of every design in engineering and some of the main benefits are explained as follows.

The principal advantage of prototyping is that it brings about a quicker and more viable design cycle. Since prototypes permit organizations to test their design in real-time environment, it is simpler to identify possible issues and forestall expensive errors not too far off. Developers can turn all the more rapidly and cooperate with the designers to investigate and further develop any problem areas.

Rapid prototyping is especially favorable on the grounds that it can create models, days or even weeks quicker than customary prototyping techniques. A wide scope of 3D printing materials, including metals and thermoplastics, is accessible, giving developers and engineers an actual portrayal of how their item will look, feel, and act.

10.2.2 AI and robotics are very crucial in healthcare systems but a lot of research is still needed

From the findings and discussion of the whole book, the following deductions were made by the authors and these should be able to help qualified personnel and trainees in the field of robotics, biomedical engineering, software engineering, mechanical engineering, mechatronics, computer science, and medicine.

From the deductions it was affirmed that almost everyone is encountering the advantages of artificial intelligence; researchers are hypothesizing the more promising time to come for artificial intelligence. However, everyone is captivated by ideas like self-driving vehicles and customized shopping; would they say there are any dangers from this tremendous part of science that we as a whole are unconscious of?

Nonetheless, when you take a look at the more splendid side of a thing, you ought to recognize that there is a hazier side to it. Likewise, regardless of the multiple benefits that artificial intelligence offers, it additionally has a few impediments that we cannot disregard. Thus, here we have explored a portion of the significant burdens of artificial intelligence adaptation that were accounted by the researchers during the above-described studies.

10.2.2.1 Exorbitant implementation cost

Developing AI-based machines, robots, and so forth involves immense costs, given the intricacy of designing that goes into creating one. Further, the cosmic cost does not stop there, as the maintenance of these systems also requires huge amounts of money. For example, Apple purchased a virtual assistant called SIRI and the procurement of the product cost somewhere near an astounding $200 million US dollars (Rao, 2010). Further, the significant expense of artificial intelligence execution is clear from the way that Amazon procured Alexa for $26 million in 2013 (Morgan, 2018).

AI-based software programs require successive upgrades so as to take into account the necessities of the changing environment, as the machine needs to become more brilliant continuously. In the event that the software experiences a serious breakdown, then, at that point, the most common way of recuperating lost codes and reinstalling the framework can be a difficult situation for anyone involved because of the immense time and cost involved.

10.2.2.2 AI and robotics systems cannot supplant humans

It is unquestionable that machines perform significantly more proficiently when contrasted with an individual. Nevertheless, and, after it's all said and done, it is for all intents and purposes difficult to supplant people with AIs, essentially sooner rather than later, in light of the fact that you cannot assemble human knowledge in a machine as it is an endowment of nature. In this way, regardless of how savvy a machine can become, it can never supplant a human.

We may get alarmed at being supplanted by machines, yet truly, it is as yet a fantastical thought. Machines are capable of reasoning yet do not have any feelings or virtues. They come up short on the capacity to bond with people which is a basic characteristic expected to deal with a group of humans. Indeed, the facts confirm that they can store a great deal of information; however, the technique of recovering data from them is a significant vexatious cycle, which is way troublesome contrasted with human knowledge.

10.2.2.3 Inability of advancement regardless of experience

One of the most astonishing qualities of human intellectual ability is its capacity to evolve with age and experience. Nonetheless, the equivalent cannot be said about AIs as they are machines that cannot improve with experience, rather it begins to mileage with time.

One needs to comprehend one thing that machines cannot modify their reactions to evolving conditions. That is the essential reason on which AIs are constructed: monotonous nature of work where the information does not change. In this way, at whatever point there is some adjustment of the input, the AIs should be reevaluated, reprepared, and reconstructed.

Machines cannot judge what is correct or not since they are unequipped for understanding the idea of ethical or legitimate. They are programmed for specific circumstances and as such cannot take choices in situations where they experience a new circumstance.

10.2.2.4 AI and robotics systems are stellar but still they are not humans

As currently referenced before, AIs are not developed for inventive bits of work. Thus it ought to be perfectly clear at this point that innovativeness or creative mind is not the specialty of the AIs. Despite the fact that they can help you in planning and making something extraordinary, they actually cannot contend with the human mind. Their inventiveness is restricted to the innovative capacity of the individual who programs and regulates them.

Human brains are described by tremendous acuteness and high sentimental feelings. To lay it out plainly, AIs can become talented machines; however, they can never obtain the capacities of the human mind. The explanation is that abilities can be acquired and mastered; however, capacities are inherently acquired and must be sharpened.

10.2.2.5 Robots might lead to unemployment

With quick improvement being made in the field of artificial intelligence, the inquiry that torment our instinctive mind is that, will AI supplant people? Truly, no one is certain about whether AIs will result in higher unemployment or not. Yet, AIs are probably going to assume control over most of the tedious undertakings, which are to a great extent simple in nature and include least subjectivity.

From the earlier discussion, it is irrefutable that a lot still needs to be done when it comes to the implementation of AI and robotics. This suffices to say, areas of research for further study can focus on finding effective solutions to the earlier challenges we just mentioned.

10.2.3 High software efficiency is everything in AI and robotics

Artificial intelligence for the most part is explored together with Machine Learning (ML) and information analytics (Aristodemou & Tietze, 2018).

ML takes information and searches for basic patterns. In the event that it spots something significant for a viable issue, software creators can take that information and use it to dissect explicit issues. Everything software is information adequately vigorous that algorithms can perceive valuable trends. Information can come as digital data, satellite pictures, visual data, text, or unstructured information.

An exhaustive research was done by the authors on AI and robotics, albeit we did not uncover everything related to subject matter but we noticed and appreciated that the main chief cornerstone of AI and robotics is the software. For example, in the context of this book, for a medical robotic manipulator to perform its intended functions, that is, retrieving and testing COVID-19 samples precisely without hurting any human being or multifunctioning, it must have a highly sophisticated software which is very efficient and secure. The following section will cover what is cybersecurity, general importance of cybersecurity, and its importance in robotic machines.

10.2.4 More advanced AI require more advanced cybersecurity systems

The robotic machine for testing COVID-19 is highly dependent on web applications and obviously the internet in general, since there is a lot of communication between the machine and the patients in real time without any human intervention. With this in mind, it is now clear that a lot of patients' confidential medical data will be stored within the machine system databases and hence they should be highly protected from cybercriminals.

As applications become more famous, mobile phones have turned into a significant objective for cybercrime. This is because a tremendous number of people now have access to smartphones. According to Bankmycell, in 2021 the total number of individuals who own smartphones is approximately 3.8 billion which is equivalent to about 48% of the world's population (Bankmycell, 2021).

The other explanation is that versatile mobile applications are currently being utilized for a vast number of exercises. A greater part of cell phone clients, for instance, really like to browse email on their mobile phones as opposed to on a computer. Individuals additionally use applications for GPS navigation and to peruse the web. Yet, most importantly, an expanding number of buyers are making procurements through applications. This implies that many individuals currently present their email addresses, financial balance numbers, Visa data, and passwords to online retailers by means of applications. Taking into account that this is actually the sort of data cybercriminals are keen on, applications are progressively seen as a more beneficial objective. In the same manner one needs to consider that: being that cell phones are compact they are exceptionally simple to steal. Therefore proprietors of lost cell phones could lose every one of the information put away in their applications. Further, most apparently real portable applications being downloaded from application stores are malevolent. Hackers could create their own applications intended to keep an eye on a user's exercises.

10.2.4.1 Cybersecurity explained

Cybersecurity is the state or cycle of securing and recovering computer frameworks, networks, gadgets, and projects from any kind of cyberattack (Yohanandhan & Elavarasan, 2020). Cyberattacks are an undeniably modern and developing risk to one's delicate information, as assailants utilize new techniques fueled by social designing and artificial intelligence to dodge customary information security controls.

The truth is that the world is progressively dependent on innovation and this dependence will proceed as we present the up-and-coming age of new technology that will approach our associated gadgets by means of Bluetooth and Wi-Fi. To keep users' information secure while accepting new information, clever cloud security arrangements ought to be executed to forestall unapproved access and urge the utilization of solid passwords.

10.2.4.2 Importance of cybersecurity

Cybersecurity's significance is on the ascent in light of the fact that it shields all classifications of information from theft and harm. This incorporates acute information, personal data, protected health data, personally recognizable data, intellectual property, information, and government and industry data frameworks. Essentially, the general public is more technologically dependent than any other time and there is no sign that this pattern will slow. Information releases that could bring about wholesale fraud are currently freely posted via online media accounts. Delicate data like federal retirement aide numbers, credit card data, and financial account details are presently put away in cloud storage administrations like Dropbox or Google Drive (Daryabar et al., 2016).

The truth is whether you are an individual, private venture, or enormous multinational, you depend on computer frameworks consistently. Pair this with the ascent in cloud administrations, poor cloud administration security, cell phones, and the Internet of Things (IoT), we have a heap of cybersecurity dangers that did not exist years and years prior. We need to comprehend the contrast among cybersecurity and data security, despite the fact that the ranges of abilities are turning out to be more comparative. This is essential, especially when it comes to the various uses of robots these days.

Without an online protection program, an association cannot shield itself against information breach crusades, making it a conspicuous objective for cybercriminals. Both inborn danger and leftover danger are expanding, driven by worldwide connectivity and use of cloud administrations, similar to Google drive, to store delicate information and individual data. Far-reaching poor design of cloud administrations matched with progressively skilled cybercriminals implies the danger that an association experiences from a fruitful digital assault or information breach is on the ascent.

Business pioneers can no longer exclusively depend on unique cybersecurity arrangements like antivirus software and firewalls, as cybercriminals are getting more astute and their strategies are turning out to be stronger to traditional cyber protections. Digital dangers can emerge out of any level of an association. The staff should be educated about straightforward social designing tricks like phishing and more refined cybersecurity assaults like ransomware assaults or other malware intended steal licensed innovation or individual information.

General Data Protection Regulation (GDPR) and different laws imply that cybersecurity is not something organizations of any size can overlook any longer. Security occurrences consistently influence organizations of various sizes and frequently make the first page, causing irreversible reputational harm to the organizations in question. To assist with understanding the significance of cybersecurity, we have first of all explained what cybersecurity is and gave some examples on various components of cybercrime most people are not aware about.

10.2.4.3 Cybersecurity a necessity for medical IoT robots

For a robot to be protected, it should likewise be secure. Cyber-physical frameworks are on the ascent. Medical 4.0, industry 4.0, and the vision of savvy, connected hospitals and plants keep on driving the advanced robotics blast.

The new advanced insurgency drove robots to become incorporated like never before into various areas like farming, healthcare, military, police, and industry. Robots are dedicated to serve, work with, and upgrade the human existence. Nonetheless, numerous episodes have been happening, prompting genuine wounds and devastating effects like the pointless loss of human lives (Alemzadeh et al., 2016). Accidental mishaps will consistently occur, yet the ones brought about by noxious assaults address an extremely difficult issue. This incorporates malignantly commandeering and controlling robots and causing genuine monetary and economic misfortunes.

Shrewd healthcare facilities are utilizing networked robots and the insightful information they create to improve the accuracy of medical surgical procedures, work on robot maintenance, amplify medical equipment production proficiency, improve the quality of hygiene in hospitals, and help reduce the risk of endangering healthcare professional lives. As more robots are associated with one another, the enterprise and the cloud, cybersecurity risks have increased.

These dangers go past information breaches, surgical procedure, and manufacturing delays. Cyberattacks focusing on exceptionally deft, amazing robotic frameworks accompany genuine well-being concerns. Robots are efficient machines determined when assigned to a mission. Therefore it is essential to make sure that their goals are all around good consistently.

With the expanding popularity of human to robot interaction and versatile robots, these worries are elevated. Artificial intelligence-empowered robots are acquiring highly advanced autonomy. The genuine strength of the present robots dwells in their intelligence, that is, the software possibly vulnerable to cyberattack.

With the assembly of Information Technology (IT) and Operational Technology (OT), network safety is now everybody's concern. In this era of the Medical Internet of Things (MIoT) and Industrial Internet of Things (IIoT), cybersecurity is everybody's obligation.

10.2.5 Using both fuzzy logic and neural networks can improve proficiency

For the purposes of presentation, the fuzzy logic controller was used to control the robotic manipulator but to improve the proficiency of the system, which is essential especially for medical robots, other categories of AI should be used together with fuzzy logic. Other categories of AI may include and are not limited to Neural Networks (NN), Genetic Algorithms (GA), Expert Systems (ES), and Neuro Fuzzy (NF) systems. For the purpose of further study in the context of this book, we will consider incorporating neural networks first and then other systems later. Therefore, following we will explain the distinct difference between fuzzy logic and neural

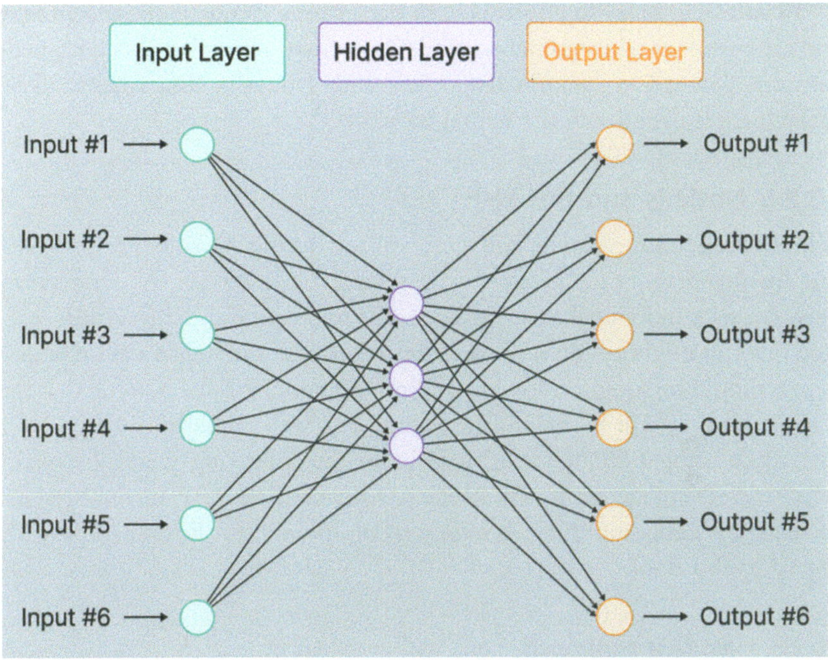

Fig. 10.1 Neural networks. *(Courtesy Marvellous Moyo and Tawanda Mushiri.)*

networks, showing clearly how neural networks can be a better choice when applied to our intelligent robotic machine for COVID-19 virus.

The principal distinction between fuzzy logic and neural network (see Fig. 10.1) is that the fuzzy logic is a thinking technique that is like human thinking and decision making, while the neural network is a framework that depends on the natural neurons of a human mind to perform calculations.

Artificial intelligence is the imitation of human insight frameworks by machines, particularly, the computer frameworks. At the end of the day, it gives a machine or a computer the capacity to perform assignments like a human. Fuzzy logic and neural network are two subclassifications of artificial intelligence. In a nutshell, these advances help to create valuable applications that can make viable choices.

10.2.5.1 Contrast between fuzzy logic and neural networks

Fuzzy logic is a thinking procedure that is similar to human decision making and manages obscure and imprecise data, while a neural network is a framework propelled by biological neurons in the human mind and can perform processing errands quicker.

In addition, fuzzy logic assists with performing trends acknowledgment and arrangement of assignments, while the neural network assists with performing perception, acknowledgment, and grouping assignments. The fuzzy logic is easier than the neural network.

10.2.6 Node.js and not PHP

A lot of programming languages were utilized during the software design of the intelligent robotic machine for testing COVID-19. A connection between each individual code was supposed to be done and the authors realized that this might result in slowed execution time and hence a better alternative should be sought. Now, to address this problem, we noticed that for further study it will be helpful to change our web application design and instead of using PHP we use Node.js. In the following we will explain why it is advantageous to use Node.js compared to PHP in our systems or related systems and thus we highly recommend it for further study with regard to this book.

10.2.6.1 Node.js explained

Node.js is neither a system nor a library. It is a cross-stage runtime environment for JavaScript (Holmes & Schur, 2021). It is a free, open-source, and full-stack development environment that can be utilized to fabricate server side and customer side of a web application. It utilizes V8, an open-source elite JavaScript engine, to design web applications. To convey superb execution, Node.js utilizes an event-driven, nonimpeding Input/Output model to fabricate information concentrated real-time applications.

Node.js is generally utilized for developing enormous scope applications. It is broadly utilized for building real-time applications, video streaming sites, single-page applications, and other web applications for different organizations.

10.2.6.2 Node.js versus PHP

Both Node.js and PHP have for quite some time been incredible choices for backend development. As on account of some other innovation, the dev community is separated with regard to PHP versus Node.js (Chaniotis, Kyriakou, & Tselikas, 2015). On one hand, we have PHP which has stayed as one of the most well-known technologies ever. On the other, we have a somewhat more up-to-date, much more current innovation Node.js that made it conceivable to utilize JavaScript for backend development.

Node.js is one of the few asynchronous development environments. This, being a significant differentiator, gives an edge over other languages. Being asynchronous implies that it does not wait for a module to be effectively executed prior to stacking the following one. Such execution radically diminishes the downtime for the web application and makes for a coherent user experience.

Like most languages and technologies from the 1990s period, PHP works on synchronicity. That implies each module and capacity is executed in the code-indicated request. On the off chance that one capacity or module is not executed, the considerable ones will not begin until it is finished. Thus, in case a backend innovation that makes an application quicker is needed, then one should definitely go for Node.js.

Since Node.js has developed with the appearance of NoSQL databases, it is very much synchronized with all types of databases. With JavaScript Object Notation, a remarkable arrangement of information utilized while connecting with data sets can work great with NoSQL databases like MongoDB and CouchDB. It additionally upholds diagram databases and has considerable libraries to access SQL databases.

Being the more established technology, PHP is intended to work with relational and customary databases like MySQL and MariaDB. While it is feasible to import libraries to work with NoSQL data sets, the cycle is monotonous and gobbles up an enormous piece of handling time.

To put it plainly, on the off chance that one intends to have a versatile web application that will often pull information from customary, social, or NoSQL databases, Node.js is the thing that one should be searching for.

The Node.js environment has a wide scope of libraries and systems; however, they fall way behind PHP's numbers. In spite of the fact that Node.js lags in quantity, it compensates for with its wide number of projects. Since it is utilized for both server-side and backend programming, the sorts of projects accessible reach unreservedly squeezing into a bigger number of utilization cases.

WordPress has been a significant contributor to the PHP environment. The organization is straightforwardly answerable for running an enormous level of complete sites on the internet, and nothing else shows PHP's heavenly reach better than this. Besides, the PHP people group has fostered a huge load of preparing material and supporting technology over the course of the years to welcome new engineers on board. Note that PHP has a bigger environment while Node.js has a more extravagant one with a more prominent assortment for projects, systems, and modules.

10.2.6.3 Benefits of using Node.js

JavaScripts are an indistinguishable piece of the web application while creating frontend applications. It is one of the generally utilized and profoundly famous customer-side programming languages. Angular.js, Node.js, and React.js fall under the top of JavaScript. Out of these three, Node.js stands firm on top of the list. Fig. 10.2 shows that a larger number of students know Node.js compared to other JavaScript programming languages (Altexsoft, 2019).

Outfitted with current tooling, Node.js permits engineers to construct complex web applications with various provisions, usefulness, and modules. This is the vital benefit of the technology that makes Node.js stand separated from other programming languages.

Utilizing the Node.js for web application development, engineers think that it is not difficult to scale the application in an upward direction and on a level plane. While scaling the application on a level plane, web designers can add new nodes to the current framework. Then again, to accomplish vertical adaptability, software engineers need to add extra resources that can be added to the current framework.

As the technology utilizes the event component, it gives high adaptability and helps the server in a nonblocking reaction. Node.js applications do not need an enormous block in the core design. It can work with a bunch of miniature administrations and modules which are running in their own

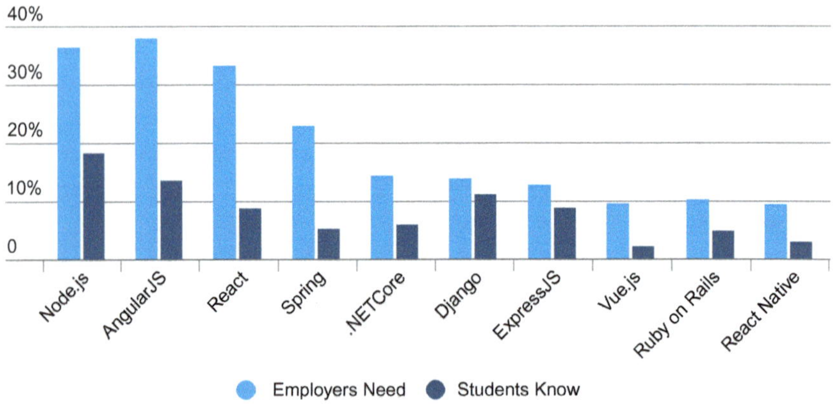

Fig. 10.2 Popularity of JS programming languages. *(Courtesy Altexsoft. (2019).* Pros and cons of Node.js web app development. *AltexSoft. Available at: https://www.altexsoft.com/blog/engineering/the-good-and-the-bad-of-node-js-web-app-development/. (Accessed 25 September 2021).)*

cycle. In this manner, it is an exceptionally adaptable environment than the other JavaScript servers.

Node.js deciphers the Java code through Google's V8 JavaScript engine. The V8 engine consents to JavaScript code straightforwardly into the machine code. It empowers engineers with the powerful execution of the code. As Node.js is upheld by the nonblocking run input-output activities, engineers can take the advantage of quicker code execution alongside an improved runtime environment.

Node.js is an exceptionally extensible technology and it very well may be altered as the changing and explicit necessities of the business and the project. Additionally, utilizing JavaScript Object Notation (JSON) with Node.js, web engineers can accomplish information trade between the customer server and web server. Software engineers can likewise utilize built-in APIs (Application Programming Interfaces) to create Transmission Control Protocol (TCP), Hypertext Transfer Protocol (HTTP), Domain Name System (DNS), and so on server.

With Node.js, designers can fabricate frontend and backend of the web applications in JavaScript utilizing a runtime environment. It empowers software engineers with the capacity to compose code for the server-side applications utilizing JavaScript. They do not have to utilize other worker-side programming languages. Consequently, application sending becomes simpler for the designers, as practically all programs support JavaScript.

Node.js, an open-source runtime environment, gives a storing module as well. At whatever point the application gets the solicitation for the main module, the solicitation is stored in the application memory. It helps designers by decreasing the assignments of the reexecution of the code. The reserving module stacks the site pages quickly and reacts to the users inside in no time.

Being a lightweight technology and upheld by the fluctuated apparatuses, Node.js permits designers to speed up the application improvement speed. The Node Package Management gives a few libraries and answers for building web applications with more worth by contributing less effort.

By offering reusable layouts, the innovation decreases the quantity of bugs, reduces the size of the application, and works on an opportunity to market of the application by shortening the designing cycle. As the customer side and server side are composed utilizing Node.js, it turns out to be simple for designers to deal with the information by devouring less planning.

Node.js has a huge and dynamic local area of software designers who persistently add to the further improvement and advancement of the technology. The web engineers are very much upheld by the JavaScript designers who give them the preparation to utilize code in GitHub to assemble their perplexing arrangements.

The technology is fit to deal with the various demands simultaneously by offering the alternative of nonblock I/O frameworks. The framework can deal with a few solicitations with the least reaction time and in an exceptionally productive manner contrasted with different languages such as Python or Ruby. Every one of the approaching solicitations get arranged and immediately executed in an exceptionally methodical manner.

JavaScript is a generally utilized programming language among the designers to fabricate the frontend of web applications. It turns out to be simple for software engineers to learn and take on Node.js for backend development. It assists them with composing code significantly quicker.

Node.js improves application reaction time, advancement time, and usefulness to empower organizations with an upper hand. The innovation holds the powerful effectiveness and flexibility by dealing with the developing volume of portable traffic, saving the server from slamming and circulating the heap. Node.js designers can foster responsive locales a lot quicker by coordinating important components.

10.2.7 Having both a web and mobile application is beneficial

To access the information of the patients, online web application forms were utilized. This was a very good method but still it has its own downsides, thus we in for further study we propose the development of both a mobile application and web application. By combing the strengths of these two types of applications, the proficiency and reliability of our systems or any systems depending on these types of software will be highly increased. Some of the advantages and disadvantages of these application platforms are discussed as follows.

10.2.7.1 Web application

To assess one's decision between a mobile website and a mobile application definitively, one has to know the advantages of mobile website just as its restrictions.

Advantages of web applications

A website improves the user experience across various kinds of cell phones. Interestingly, a mobile application requires fostering a different form for each working framework and gadget type. Users who own gadgets of various kinds may particularly see the value in the advantages of similarity that responsive websites give. Also, they support simple reconciliation with other portable elements like Quick Response (QR) codes and text messaging.

Due to the multigadget support that responsive website architecture gives across different stages, it is turning out to be not difficult to contact a more extensive crowd than a mobile application can reach.

Websites likewise cost less to update. You need to keep up with simply a solitary version of them. Contrasted with mobile applications, which require downloading of each and every update, responsive or mobile websites let you change the substance or plan just by altering them once, and permit you to do that effectively and deftly. In the wake of executing, refreshes become dynamic and noticeable promptly across a wide range of gadgets.

Disadvantages of web applications

Mobile websites have a more extensive reach, better similarity, and require less resources than versatile applications. In any case, they have their limits. In contrast to an application, a responsive/mobile website cannot use all cell phone features as effectively. Cameras, Global Positioning System (GPS), telephone dialing, and different components coordinated into cell phones are not in every case very much created for responsive/mobile websites, despite the fact that APIs and libraries expected to assist with settling these issues have been showing up of late.

One more restriction to user benefits is a gadget's screen size. Obviously, portability is the key motivation behind why users appreciate having more modest gadgets in a hurry. Nonetheless, cell phones show significantly less substance contrasted with a monitor or laptop screen. The responsive design and mobile website do not completely have access to all the contents accessible on the desktop. Despite the fact that responsive website composition acclimates to the screen size dynamically, still it does not make it totally helpful for users to ride the content on a cell phone. Exactly the same thing is with mobile website which really decreases and revises the substance accessible on the desktop.

Since mobile encounters altogether vary from desktop encounters, executing similar interface for the two stages might hurt one's User Experience (UX) technique. This for the most part identifies with the single-window limitation. The single-window limitation says that a user should not pass on their present page to get to all the content they are searching for; in any case, with mobile sites it is usually difficult to fit everything onto one page.

Regardless of whether you plan your mobile website in the lightest and most educational manner conceivable, it actually may work disconnected with just restricted usefulness utilizing stored pages. In contrast to a mobile application, which can run locally, a mobile website requires a decent internet connection to completely.

10.2.7.2 Mobile applications

Obviously, to choose whether you need to develop an application, it is essential to explore the critical benefits of mobile applications just as their downsides.

Advantages of mobile applications

Investigation shows that the applications are more well known than identical websites, as they are more helpful (Deshdeep, 2021). Mobile applications give better user encounters, load content quicker, and are simpler to utilize. Additionally, in contrast to websites, applications have pop-up messages. Sharing updates, exceptional elements, and updates inside an application builds user dedication and maintenance. Additionally, the design of mobile applications fits diverse screen estimates more exquisitely than sites.

Mobile applications are an incredible answer for administrations that require normal use. An application permits users to set inclinations, make individual records, and keep imperative data nearby. According to a business perspective, mobile applications offer better help for focusing on a crowd of people and thus creating marketing campaigns for various gatherings of users.

Working disconnected. One more significant benefit identified with mobile applications is the chance to utilize them disconnected. As applications are introduced on a cell phone, they can continue to give admittance to content and elements even without an internet connection.

Disadvantages of mobile applications

To guarantee legitimate working, a mobile application should meet the prerequisites of the specific working framework. This implies that each platform, iOS, Android, and Windows, requires a different application version.

Obviously, it is feasible to foster a cross-stage arrangement that should fit each operating system and gadget. Nonetheless, cross-platform application structures do not offer full help for each module and component of each working framework.

At the point when an application is produced for multiple unique platforms, supporting it likewise takes additional time and resources. Without a doubt, you need to give overhauls and fix similarity issues for each kind of gadget and do as such consistently. Moreover, you need to advise your users about these updates and push them to download the refreshed form of the application. One more issue identified with keeping an application is the time and exertion expected to get endorsement from the business sectors where the application is put.

10.2.8 Use of robotic simulation tools

Limitations were encountered during the project which include time and necessary resources like software. As mentioned before, prototypes were not developed mainly due to lack of resources and they are very essential, especially for medical robots. Therefore we recommend use of robotic simulation tools in further studies as this will help virtualize how the robot will be operating in the real world. So, what are robotic simulation tools?

Robotic simulation tool is developing robotic automation as a practical alternative for almost any manufacture. It is utilized during the citing and concept stage to give confirmation of configuration, evidence of interaction, and to amplify a robot user's automation investment. Examples of robotic simulation tools are SimSpark and Webots.

There are many reasons why most robotic integrators utilize some type of robotic simulation tool in their projects. However, what are the advantages to producers? For what reason should a producer be keen on robot simulation?

10.2.8.1 Importance and benefits of robotic simulation tools

Basically, robotic simulation tool reduces the dangers related with automation. A large part of the joining system is finished prior to anything is collected on your industrial facility floor. In the context of this book, these will be the number of patients tested by the robot and specimen it will be analyzing. An appropriate simulation guarantees that one is getting the best framework for their application.

Processing capacity is continually improving, just as becoming less expensive, making robot simulation increasingly more essential to discovering a framework that will assist one to stay competitive in their line of business.

A manufacturer benefits in many ways from advanced robot simulation software. However, there are three main ways a manufacturer directly and immediately benefits:

A biomedical engineer developing and manufacturing medical robots benefits in numerous ways from cutting-edge robotic simulation tools. In any case, some of the principal ways a producer will straightforwardly and promptly benefit are discussed as follows.

Frequently, medical robotic frameworks are not one-size-fits-all. Robotic simulation tools assist perfect and customize robotic design to guarantee the framework will fit consistently with existing systems and equipment in hospitals. Medical robotic manipulators should be extremely safe because they interact directly with human beings.

Robotic simulation tools can demonstrate the adequacy of an automation framework. With exact, point-by-point simulation one can be certain that an automated framework will meet one's process duration and effectiveness objectives. Without this simulation, one will be facing a major challenge that the proposed framework will work effectively for their specific application.

When a simulation is finished, the outcome is ordinarily downloadable software program for one's robotic application, so the genuine framework will coordinate with the simulation precisely. Robotic simulation tools guarantee potential imperfections are not planned into the framework and one's robotic framework works unequivocally as expected. This will make sure that a high number of COVID-19 tests can be done without any harm coming to the patients, while at the same time minimizing the infection risks of healthcare workers.

10.2.9 Life cycle analysis

Pandemics like COVID-19 do not care about social status, ethnic group, race, and so on, thus the WHO has termed it as a global pandemic. Therefore it is up to the whole world to work together and addressing such problems will not be much of an issue to handle in the future. Unfortunately, Africa is still lagging behind other continents when it comes to almost everything, especially technology, though most of the crucial things found in the first world countries are now available in Africa as well, like IoT. Thus companies that develop similar systems should also take into account where possible that they should include Africa as well. This is because a lot of challenges were encountered during the life cycle analysis study such as the unavailability of the software use in Africa and hence future software designs can be done that include Africa as well, as this subject matter is crucial during the life of any product in any continent.

10.3 Future work

The proposed manipulator was supposed to be fixed at the base on the table and this was advantageous since the design will be simple and minimum number of materials will be used. Yet the authors for future work recommend installing the robotic manipulator hanging upside down from the ceiling of the booth instead of fixing it on the table and using three-axis prismatic joints to move it in the x, y, z directions. This will help improve the dexterity of the robot during testing and hence its efficiency. This future work can go on further to compare the costs and effectiveness of these two positions of the robotic manipulator, with the main objective being to select the best option for final production of the intelligent robotic machine for testing COVID-19 virus.

A few unique sensations can be felt during PCR testing for COVID-19 virus (Koenig, 2021). Individuals have announced transient torment, profound burning inside the nose, choking when the rear of the throat is contacted, sneezing, coughing, and tearing because of the setting off of a nasal lacrimal reflex. On the off chance that you end up having a critical septal deviation or have obstructed nasal entries because of sensitivities, cold-like manifestations, and additionally nasal polyps, it very well might be especially difficult to pass the nasal swab to the rear of the throat. Making system of gathering nasal discharges through a nasopharyngeal swab simpler for the patient might build better testing acceptance and consider better quality examples for testing. Therefore future work should focus on the design of systems that can retrieve COVID-19 samples without causing the patient too much discomfort.

10.4 Conclusion

The ultimate goal of this research was to reduce the rate of COVID-19 infection, especially for healthcare workers who interact daily with potential cases, by designing an intelligent robotic machine for testing COVID-19 virus. An in-depth literature survey was done on the application of artificial intelligence and robotics in the healthcare sector. To better understand the efficiency of the application of robotics and artificial intelligence in healthcare sector, the research was divided into 10 individual parts and each section was exhaustively done. The research that has be done so far covers the application of robotics and artificial intelligence that has not yet been covered in the field of medicine. According to the results, such a system will be very helpful in the reduction of COVID-19 virus infection and hence number of deaths; in addition, the same system can be easily adapted to address similar pandemics in future. Owing to this, it is irrefutable that further pursuit of this research is of paramount importance.

References

Alemzadeh, H., et al. (2016). Adverse events in robotic surgery: A retrospective study of 14 years of FDA data. *PLoS ONE*, *11*(4). https://doi.org/10.1371/journal.pone.0151470. Public Library of Science.

Altexsoft. (2019). *Pros and cons of Node.js web app development*. AltexSoft. Available at: https://www.altexsoft.com/blog/engineering/the-good-and-the-bad-of-node-js-web-app-development/ (Accessed 25 September 2021).

Aristodemou, L., & Tietze, F. (2018). Property Analytics (IPA): A literature review on artificial intelligence, machine learning and deep learning methods for analysing intellectual property (IP) data. *World Patent Information*. Elsevier. Available at: https://www.sciencedirect.com/science/article/pii/S0172219018300103 (Accessed 25 September 2021).

Bankmycell. (2021). *Bankmycell, stats of cell phones*. Google Search. Available at: https://www.google.com/search?q=+Bankmycell%2C+stats+of+cell+phones&client=firefox-b-d&bih=615&biw=1366&hl=en&sxsrf=AOaemvLS-TvHlIoSkPuQGutyAQlD_wFkQw%3A1632552894275&ei=vsdOYd6gENKV8gKZ863oCg&oq=+Bankmycell%2C+stats+of+cell+phones&gs_lcp=Cgdnd3Mtd2l6EAMyBwgAEEcQsAMyBwgAEEcQsAMyBwgAEEcQsAMyBwgAEEcQsAMyBwgAEEcQsAMyBwgAEEcQsAMyBwgAEEcQsAMyBwgAEEcQsANKBAhBGABQgvoNWIyYD2Dvnw9oAHADeAOAAQCIAQCSAQCYAQCgAQHIAQjAAQE&sclient=gws-wiz&ved=0ahUKEwje99u6xZnzAhXSilwKHZl5C60Q4dUDCA0&uact=5 (Accessed 25 September 2021).

Chaniotis, I. K., Kyriakou, K. I. D., & Tselikas, N. D. (2015). Is Node.js a viable option for building modern web applications? A performance evaluation study. *Computing*, *97*(10), 1023–1044. https://doi.org/10.1007/s00607-014-0394-9. Springer-Verlag Wien.

Daryabar, F., et al. (2016). Forensic investigation of OneDrive, Box, GoogleDrive and Dropbox applications on Android and iOS devices. *Australian Journal of Forensic Sciences*, *48*(6), 615–642. https://doi.org/10.1080/00450618.2015.1110620. Taylor and Francis Ltd.

Deshdeep, N. (2021). *App or website? 10 reasons why apps are better*. VWO. Available at: https://vwo.com/blog/10-reasons-mobile-apps-are-better/ (Accessed 25 September 2021).

Holmes, K., & Schur, A. (2021). *An evaluation of tools for identifying vulnerabilities in open source dependencies for NodeJS applications* (pp. 330–335). https://doi.org/10.1007/978-3-030-68017-6_49.

Koenig, D. (2021). *Brain scraper: Why do some COVID tests hurt so much?*. Available at: https://www.webmd.com/lung/news/20210330/brain-scraper-why-do-some-covid-tests-hurt-so-much (Accessed 25 September 2021).

Morgan, B. (2018). *How Amazon has reorganized around artificial intelligence and machine learning*. Available at: https://www.forbes.com/sites/blakemorgan/2018/07/16/how-amazon-has-re-organized-around-artificial-intelligence-and-machine-learning/?sh=44ad76937361 (Accessed 25 September 2021).

Rao, L. (2010). *Confirmed: Apple buys virtual personal assistant startup Siri*. TechCrunch. Available at: https://techcrunch.com/2010/04/28/apple-buys-virtual-personal-assistant-startup-siri/ (Accessed 25 September 2021).

WHO. (2021). *The impact of COVID-19 on global health goals*. Available at: https://www.who.int/news-room/spotlight/the-impact-of-covid-19-on-global-health-goals (Accessed 25 September 2021).

Yohanandhan, R., & Elavarasan, R. (2020). *Cyber-physical power system (CPPS): A review on modeling, simulation, and analysis with cyber security applications*. ieeexplore.ieee.org. Available at: https://ieeexplore.ieee.org/abstract/document/9167203/ (Accessed 25 September 2021).

APPENDIX A

COVID-19 testing survey questionnaire

Subject: Health professionals' perspectives on COVID-19 testing survey
Header: Design of an intelligent robotic machine for testing COVID-19 virus: Case for Mpilo Central Hospital
Introduction: **Dear Participant** Thank you for taking part in this study to help improve actions taken in response to COVID-19 pandemic and to inform the response to similar future outbreaks. This study will involve answering a 5-minute survey which will be asking you questions relating to the coronavirus. Please do not start until you will have enough time to complete it in one go. Please close other programs (e.g. chat or email) to avoid distractions. This study is conducted by Marvellous Moyo, a fourth-year student at the University of Zimbabwe, contact details: 00263 77 540 7000, email: mayezi28@yahoo.com. By taking part, you are agreeing that you have read and understood the information about the study below. Please ensure you have read and understood this information before continuing. **What is this project about and do I have to take part?** This study aims to find mitigation measures to challenges health facility are facing during COVID-19 pandemic and the main focus is to design an intelligent robotic machine for testing COVID-19 virus. The information collected through this survey is important to support the implementation of this robotic system. Participation is open to any hospital or health facility staff member. You do not have to be in isolation to take part.

What are the benefits and risks of taking part?
You may benefit from taking part in the survey by being motivated to look up information about the coronavirus pandemic. There are no foreseeable risks for you when taking part in the survey other than time spent on the survey and potential discomfort. Should you feel uncomfortable and want to leave the study you are free to do so without any consequences.

What will you ask and what will happen to the information I give you?
You will be asked questions about yourself, your knowledge of the coronavirus, the effects of COVID-19 pandemic in hospitals, and types of COVID-19 virus testing. However, you will not be asked to provide any personal data. Your anonymous data will be collected and analyzed by Marvellous Moyo. Your data will be shared, but only with relevant researchers. However, your data will be completely anonymous, and it will not be possible to identify you individually from your answers. This study has received approval from the University of Zimbabwe and has been adapted from the WHO standard questionnaire program.

Concerns
If you are concerned about this study, or how your data is being processed, or if you would like to contact us about your rights, please get in touch with Marvellous Moyo in the first instance at mayezi28@yahoo.com.

Consent
I understand that:

- My participation is completely voluntary.
- All my answers will be used for scientific research to improve actions taken in response to the coronavirus pandemic and to inform the response to similar future outbreaks.
- My data will be stored securely; however, no personal data will be stored, and my answer will be completely anonymous.
- My data gathered in this study will be shared with relevant researchers.
- Because I am submitting anonymous data, it will not be possible to withdraw my answers after they have been submitted.

Please note that you can stop the survey at any time. This will not entail any penalty, and it will not affect the services (healthcare services or others) that you receive.

By ticking the box, you are agreeing that you work or are part of any healthcare system, that you have read the information about the study, and that you voluntarily agree to take part in it.

[ ] I agree to participate in this study.

Variable: sociodemographic	**1.** What is your sex? [ ] Male [ ] Female [ ] Other **2.** What is your field of study or profession? .... **3.** What is your academic level? [ ] undergraduate [ ] postgraduate [ ] doctorate [ ] professor **4.** How long have you worked at Mpilo Central Hospital? [ ] Less than 5 years [ ] 5–10 years [ ] Above 10 years
Variable: Probability and Severity	**5.** What do you consider to be your own probability of getting infected with COVID-19? Extremely unlikely [ ] 1 [ ] 2 [ ] 3 [ ] 4 [ ] 5 Extremely likely
Variable: COVID-19 personal experience	**6.** Can anyone be tested for COVID-19 virus? [ ] Yes [ ] No **7.** If the answer to the above question is NO please elaborate on this No because: [ ] One needs an approval from the doctor [ ] One needs to show signs of symptoms first [ ] There is limited equipment to test everyone [ ] There is limited workforce to test every one **8.** Have you been tested for COVID-19 virus? [ ] Yes [ ] No

	9. How long does it take to get COVID-19 test results? [ ] 0–15 minutes [ ] 30–60 minutes [ ] 1–2 hrs [ ] 2–5 hrs [ ] 1 day [ ] More than 1 day
Variable: COVID-19 testing	**10.** What type of COVID-19 testing technique are you using? (tick as many as possible) [ ] RT-PCR Diagnosis test [ ] Antigen Diagnosis test [ ] Antibody test **11.** From your experience what is the accuracy of each test RT-PCR Diagnosis test Very inaccurate [ ] 1 [ ] 2 [ ] 3 [ ] 4 [ ] 5 Very accurate Antigen Diagnosis test Very inaccurate [ ] 1 [ ] 2 [ ] 3 [ ] 4 [ ] 5 Very accurate Antibody test Very inaccurate [ ] 1 [ ] 2 [ ] 3 [ ] 4 [ ] 5 Very accurate **12.** How many COVID-19 tests are being done per day? [ ] Less than 100 [ ] 100 to 500 [ ] 500 to 1000 [ ] Greater than 1000 **13.** Can you be admitted in the hospital for having COVID-19 symptoms? [ ] Yes [ ] No

Variable: COVID-19 effects	**14.** To what extent has COVID-19 pandemic affected the hospital? Extremely high [ ] 1 [ ] 2 [ ] 3 [ ] 4 [ ] 5 Extremely low **15.** What are the effects of COVID-19 at the hospital? (select as many as possible) [ ] Shortages of staff [ ] Inadequate personal protective equipment [ ] Shortages of beds [ ] Inadequate supply of sanitizers If more specify: ....
Variable: Medical robots	**16.** Are you aware of the use of robots to assist medical doctors in human surgical procedures? [ ] Yes [ ] No **17.** Do you know what telemedicine is? [ ] Yes [ ] No
Debriefing	Thank you very much! Your participation provides valuable insights for all of us to react appropriately in the current COVID-19 situation and to reach all citizens with useful information in a timely manner. If you have changed your opinion and would like to withdraw your consent to use your data, please click on "Withdraw my consent." This will not entail any penalty, and it will not affect the services (healthcare services or others) that you receive. [ ] Withdraw my consent

APPENDIX B

Motor sizes

MG996R servo specifications were obtained in the following link:

- https://components101.com/motors/mg996r-servo-motor-datasheet

SG90 digital micro specifications were obtained in the following link:

- https://components101.com/motors/servo-motor-basics-pinout-datasheet

MG996R servo specifications	
Modulation:	Digital
Torque:	4.8 V: 130.54 oz in. (9.40 kg cm)
	6.0 V: 152.76 oz in. (11.00 kg cm)
Speed:	4.8 V: 0.19 s/60 degrees
	6.0 V: 0.15 s/60 degrees
Weight:	1.94 oz (55.0 g)
Dimensions:	Length: 1.60 in. (40.7 mm)
	Width: 0.78 in. (19.7 mm)
	Height: 1.69 in. (42.9 mm)
Motor Type:	Brushed Motor
Gear Type:	Metal
Rotation/Support:	Dual bearings
Rotational Range:	0-180 Degrees
Pulse Cycle:	1 ms
Pulse Width:	500 ms-2500 ms
Connector Type:	JR

Specifications: SG90 digital servo
Weight: 9 g
Dimension: 23 × 12.2 × 29 mm
Stall torque: 1.8 kg/cm (4.8 V)
Gear type: POM gear set
Operating speed: 0.1 s/60 degrees (4.8 V)
Operating voltage: 4.8 V
Temperature range: 0–55°C
Dead band width: 1 µs
Power Supply: Through external adapter
Servo wire length: 25 cm
Servo Plug: JR (Fits JR and Futaba)

APPENDIX C

Engineering drawings

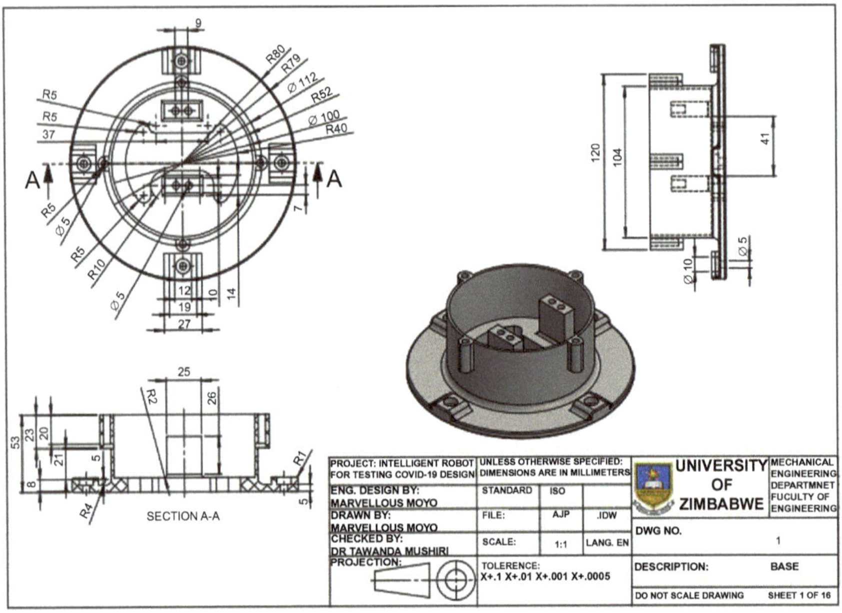
SECTION A-A
PROJECT: INTELLIGENT ROBOT FOR TESTING COVID-19 DESIGN
UNLESS OTHERWISE SPECIFIED: DIMENSIONS ARE IN MILLIMETERS
UNIVERSITY OF ZIMBABWE
MECHANICAL ENGINEERING DEPARTMNET FUCULTY OF ENGINEERING
ENG. DESIGN BY: MARVELLOUS MOYO
STANDARD ISO
DRAWN BY: MARVELLOUS MOYO
FILE: AJP .IDW
DWG NO. 1
CHECKED BY: DR TAWANDA MUSHIRI
SCALE: 1:1 LANG. EN
PROJECTION:
TOLERENCE: X+.1 X+.01 X+.001 X+.0005
DESCRIPTION: BASE
DO NOT SCALE DRAWING
SHEET 1 OF 16

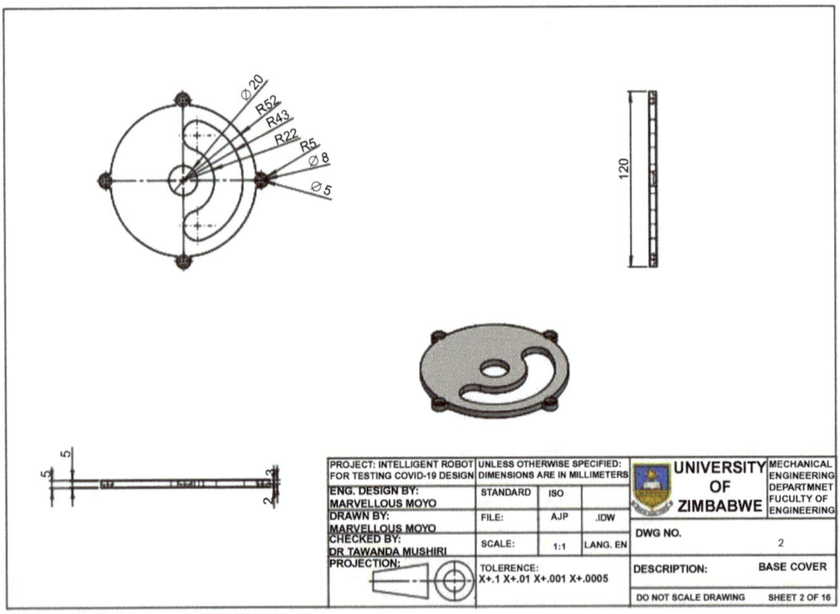
PROJECT: INTELLIGENT ROBOT FOR TESTING COVID-19 DESIGN
UNLESS OTHERWISE SPECIFIED: DIMENSIONS ARE IN MILLIMETERS
UNIVERSITY OF ZIMBABWE
MECHANICAL ENGINEERING DEPARTMNET FUCULTY OF ENGINEERING
ENG. DESIGN BY: MARVELLOUS MOYO
STANDARD ISO
DRAWN BY: MARVELLOUS MOYO
FILE: AJP .IDW
DWG NO. 2
CHECKED BY: DR TAWANDA MUSHIRI
SCALE: 1:1 LANG. EN
PROJECTION:
TOLERENCE: X+.1 X+.01 X+.001 X+.0005
DESCRIPTION: BASE COVER
DO NOT SCALE DRAWING
SHEET 2 OF 16

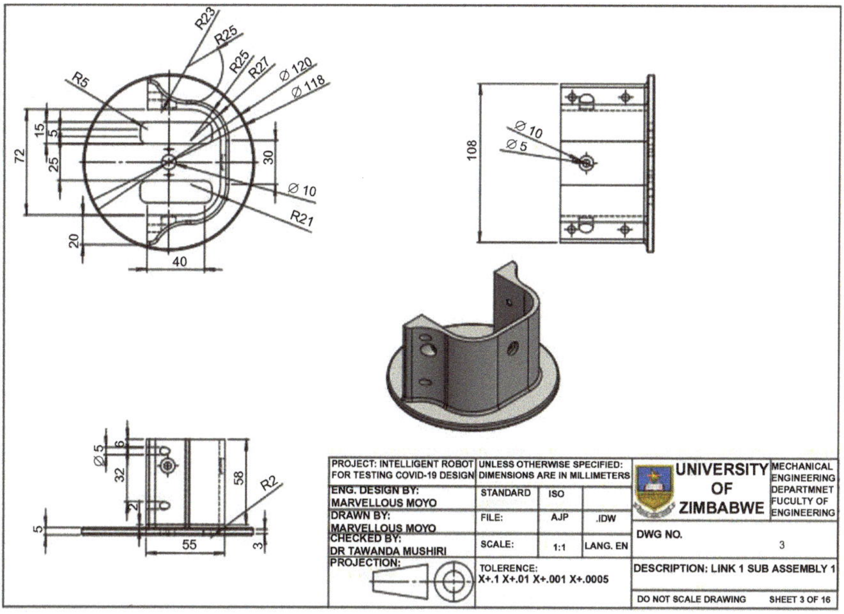
R23
R25
R25
R27
Ø 120
Ø 118
R5
15
5
72
25
30
Ø 10
R21
20
40
108
Ø 10
Ø 5
Ø 5
6
32
58
R2
2
5
55
3
PROJECT: INTELLIGENT ROBOT FOR TESTING COVID-19 DESIGN
UNLESS OTHERWISE SPECIFIED: DIMENSIONS ARE IN MILLIMETERS
UNIVERSITY OF ZIMBABWE
MECHANICAL ENGINEERING DEPARTMNET FUCULTY OF ENGINEERING
ENG. DESIGN BY: MARVELLOUS MOYO
STANDARD
ISO
DRAWN BY: MARVELLOUS MOYO
FILE:
AJP
.IDW
DWG NO.
3
CHECKED BY: DR TAWANDA MUSHIRI
SCALE:
1:1
LANG. EN
PROJECTION:
TOLERENCE: X+.1 X+.01 X+.001 X+.0005
DESCRIPTION: LINK 1 SUB ASSEMBLY 1
DO NOT SCALE DRAWING
SHEET 3 OF 16

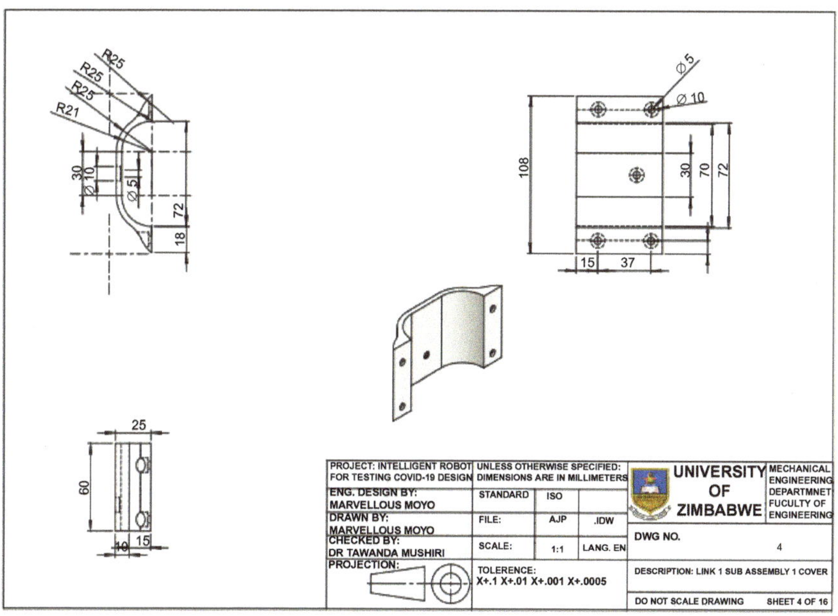
R25
R25
R25
R21
30
Ø 10
Ø 5
72
18
Ø 5
Ø 10
108
30
70
72
15
37
25
60
10
15
PROJECT: INTELLIGENT ROBOT FOR TESTING COVID-19 DESIGN
UNLESS OTHERWISE SPECIFIED: DIMENSIONS ARE IN MILLIMETERS
UNIVERSITY OF ZIMBABWE
MECHANICAL ENGINEERING DEPARTMNET FUCULTY OF ENGINEERING
ENG. DESIGN BY: MARVELLOUS MOYO
STANDARD
ISO
DRAWN BY: MARVELLOUS MOYO
FILE:
AJP
.IDW
DWG NO.
4
CHECKED BY: DR TAWANDA MUSHIRI
SCALE:
1:1
LANG. EN
PROJECTION:
TOLERENCE: X+.1 X+.01 X+.001 X+.0005
DESCRIPTION: LINK 1 SUB ASSEMBLY 1 COVER
DO NOT SCALE DRAWING
SHEET 4 OF 16

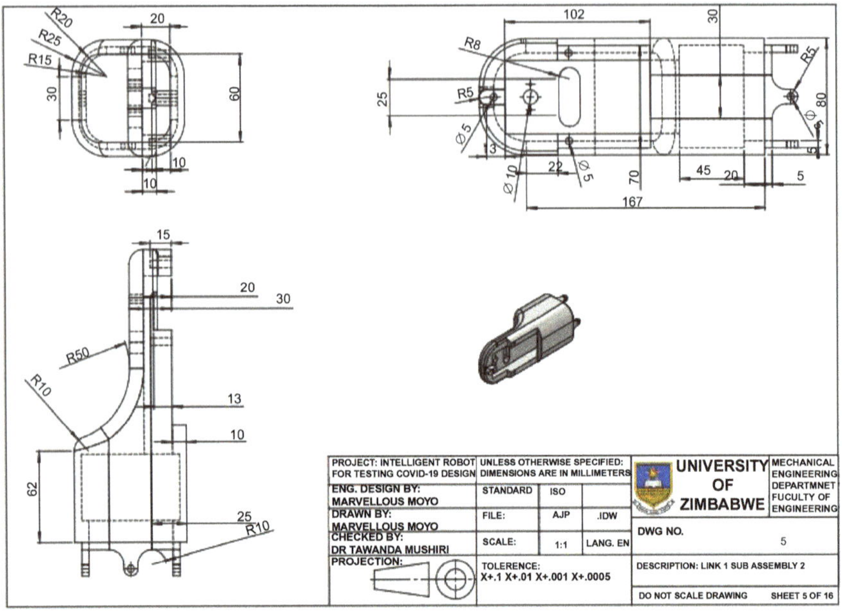
PROJECT: INTELLIGENT ROBOT FOR TESTING COVID-19 DESIGN
UNLESS OTHERWISE SPECIFIED: DIMENSIONS ARE IN MILLIMETERS
UNIVERSITY OF ZIMBABWE
MECHANICAL ENGINEERING DEPARTMNET FUCULTY OF ENGINEERING
ENG. DESIGN BY: MARVELLOUS MOYO
STANDARD ISO
DRAWN BY: MARVELLOUS MOYO
FILE: AJP .IDW
DWG NO. 5
CHECKED BY: DR TAWANDA MUSHIRI
SCALE: 1:1 LANG. EN
PROJECTION:
TOLERENCE: X+.1 X+.01 X+.001 X+.0005
DESCRIPTION: LINK 1 SUB ASSEMBLY 2
DO NOT SCALE DRAWING
SHEET 5 OF 16

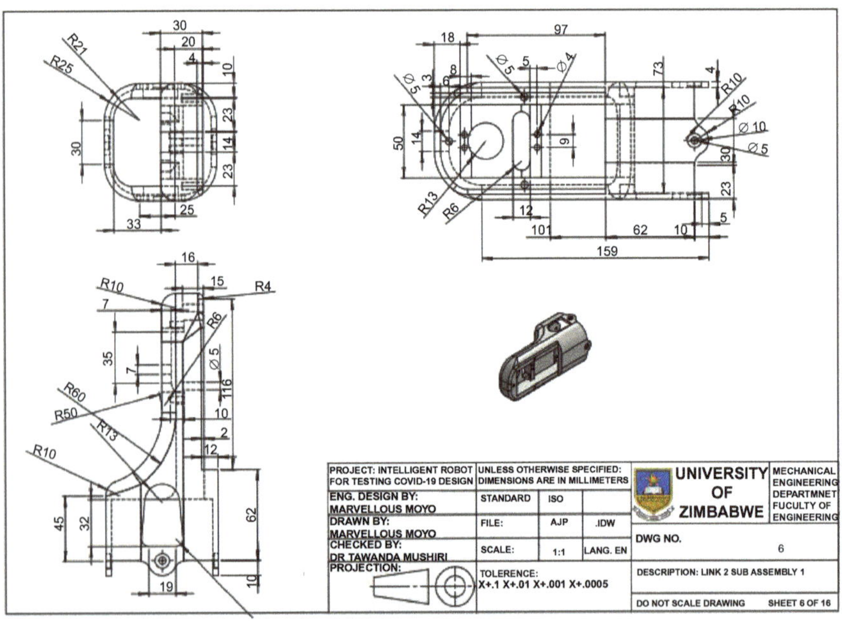
PROJECT: INTELLIGENT ROBOT FOR TESTING COVID-19 DESIGN
UNLESS OTHERWISE SPECIFIED: DIMENSIONS ARE IN MILLIMETERS
UNIVERSITY OF ZIMBABWE
MECHANICAL ENGINEERING DEPARTMNET FUCULTY OF ENGINEERING
ENG. DESIGN BY: MARVELLOUS MOYO
STANDARD ISO
DRAWN BY: MARVELLOUS MOYO
FILE: AJP .IDW
DWG NO. 6
CHECKED BY: DR TAWANDA MUSHIRI
SCALE: 1:1 LANG. EN
PROJECTION:
TOLERENCE: X+.1 X+.01 X+.001 X+.0005
DESCRIPTION: LINK 2 SUB ASSEMBLY 1
DO NOT SCALE DRAWING
SHEET 6 OF 16

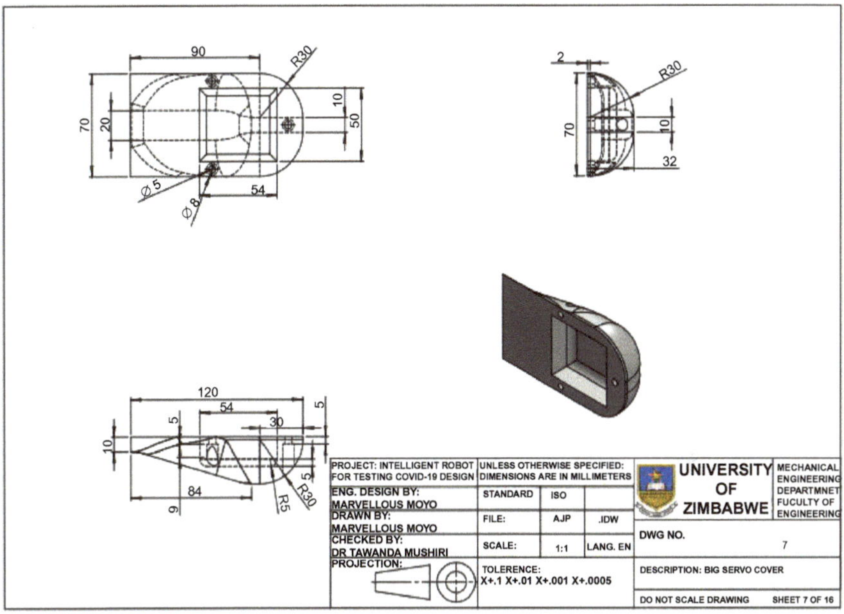
PROJECT: INTELLIGENT ROBOT FOR TESTING COVID-19 DESIGN
UNLESS OTHERWISE SPECIFIED: DIMENSIONS ARE IN MILLIMETERS
UNIVERSITY OF ZIMBABWE
MECHANICAL ENGINEERING DEPARTMNET FUCULTY OF ENGINEERING
ENG. DESIGN BY: MARVELLOUS MOYO
STANDARD
ISO
DRAWN BY: MARVELLOUS MOYO
FILE:
AJP
.IDW
DWG NO.
7
CHECKED BY: DR TAWANDA MUSHIRI
SCALE:
1:1
LANG. EN
PROJECTION:
TOLERENCE: X+.1 X+.01 X+.001 X+.0005
DESCRIPTION: BIG SERVO COVER
DO NOT SCALE DRAWING
SHEET 7 OF 16

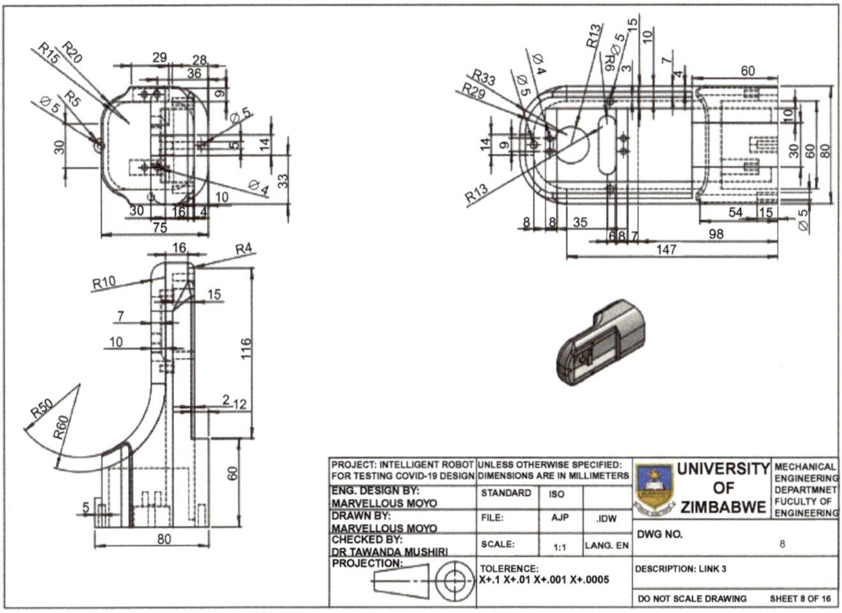
PROJECT: INTELLIGENT ROBOT FOR TESTING COVID-19 DESIGN
UNLESS OTHERWISE SPECIFIED: DIMENSIONS ARE IN MILLIMETERS
UNIVERSITY OF ZIMBABWE
MECHANICAL ENGINEERING DEPARTMNET FUCULTY OF ENGINEERING
ENG. DESIGN BY: MARVELLOUS MOYO
STANDARD
ISO
DRAWN BY: MARVELLOUS MOYO
FILE:
AJP
.IDW
DWG NO.
8
CHECKED BY: DR TAWANDA MUSHIRI
SCALE:
1:1
LANG. EN
PROJECTION:
TOLERENCE: X+.1 X+.01 X+.001 X+.0005
DESCRIPTION: LINK 3
DO NOT SCALE DRAWING
SHEET 8 OF 16

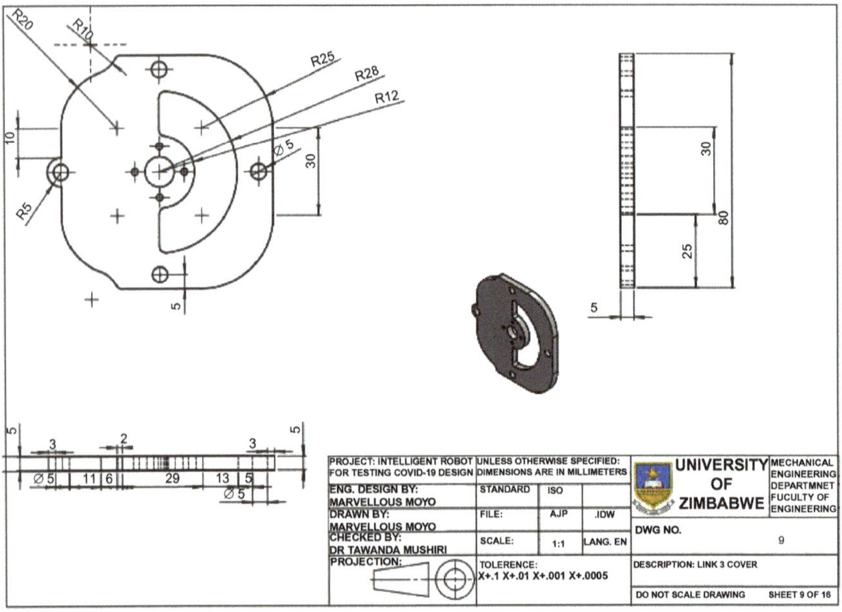
PROJECT: INTELLIGENT ROBOT FOR TESTING COVID-19 DESIGN
UNLESS OTHERWISE SPECIFIED: DIMENSIONS ARE IN MILLIMETERS
UNIVERSITY OF ZIMBABWE
MECHANICAL ENGINEERING DEPARTMNET FUCULTY OF ENGINEERING
ENG. DESIGN BY: MARVELLOUS MOYO
STANDARD
ISO
DRAWN BY: MARVELLOUS MOYO
FILE:
AJP
.IDW
DWG NO.
9
CHECKED BY: DR TAWANDA MUSHIRI
SCALE:
1:1
LANG. EN
PROJECTION:
TOLERENCE: X+.1 X+.01 X+.001 X+.0005
DESCRIPTION: LINK 3 COVER
DO NOT SCALE DRAWING
SHEET 9 OF 16

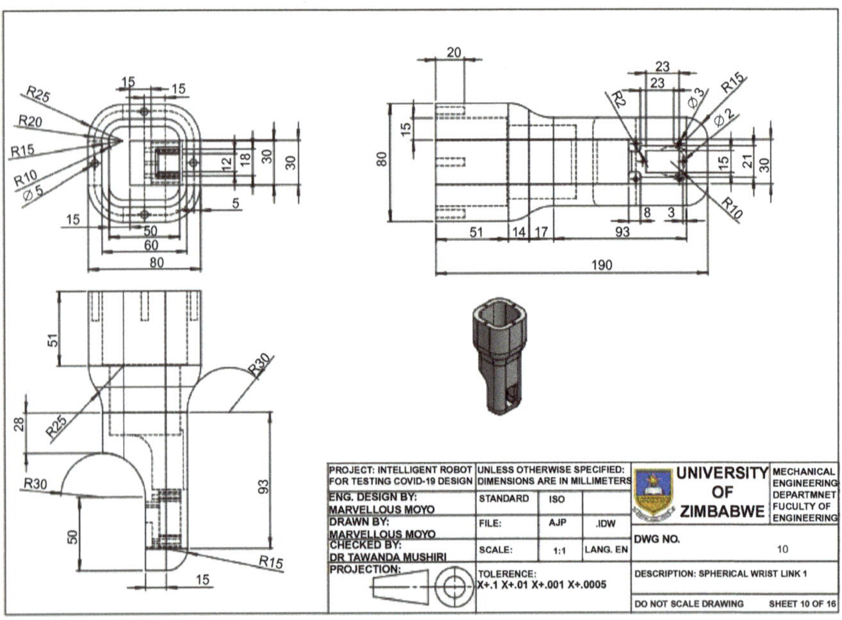
PROJECT: INTELLIGENT ROBOT FOR TESTING COVID-19 DESIGN
UNLESS OTHERWISE SPECIFIED: DIMENSIONS ARE IN MILLIMETERS
UNIVERSITY OF ZIMBABWE
MECHANICAL ENGINEERING DEPARTMNET FUCULTY OF ENGINEERING
ENG. DESIGN BY: MARVELLOUS MOYO
STANDARD
ISO
DRAWN BY: MARVELLOUS MOYO
FILE:
AJP
.IDW
DWG NO.
10
CHECKED BY: DR TAWANDA MUSHIRI
SCALE:
1:1
LANG. EN
PROJECTION:
TOLERENCE: X+.1 X+.01 X+.001 X+.0005
DESCRIPTION: SPHERICAL WRIST LINK 1
DO NOT SCALE DRAWING
SHEET 10 OF 16

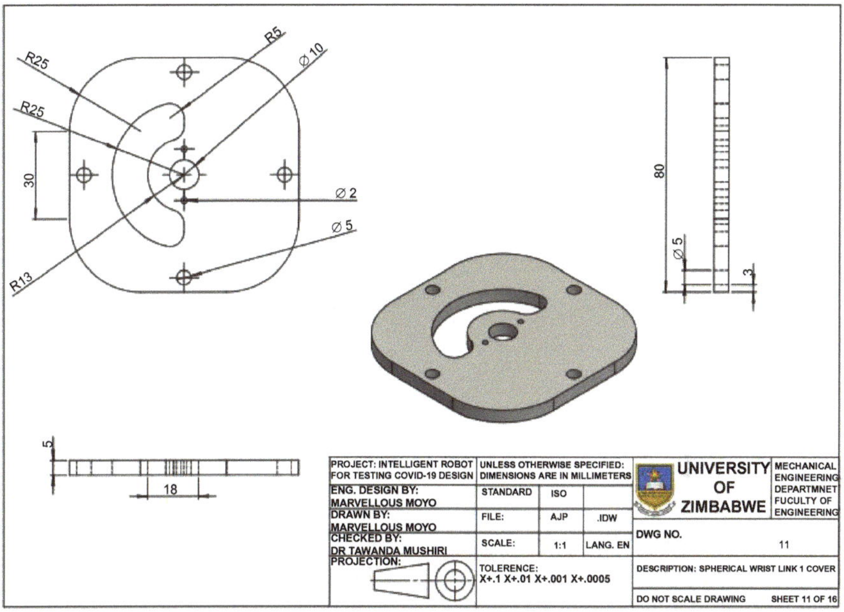
R25
R25
R5
Ø 10
30
Ø 2
Ø 5
R13
80
Ø 5
3
5
18
PROJECT: INTELLIGENT ROBOT FOR TESTING COVID-19 DESIGN
UNLESS OTHERWISE SPECIFIED: DIMENSIONS ARE IN MILLIMETERS
UNIVERSITY OF ZIMBABWE
MECHANICAL ENGINEERING DEPARTMNET FUCULTY OF ENGINEERING
ENG. DESIGN BY: MARVELLOUS MOYO
STANDARD
ISO
DRAWN BY: MARVELLOUS MOYO
FILE:
AJP
.IDW
DWG NO.
11
CHECKED BY: DR TAWANDA MUSHIRI
SCALE:
1:1
LANG. EN
PROJECTION:
TOLERENCE: X+.1 X+.01 X+.001 X+.0005
DESCRIPTION: SPHERICAL WRIST LINK 1 COVER
DO NOT SCALE DRAWING
SHEET 11 OF 16

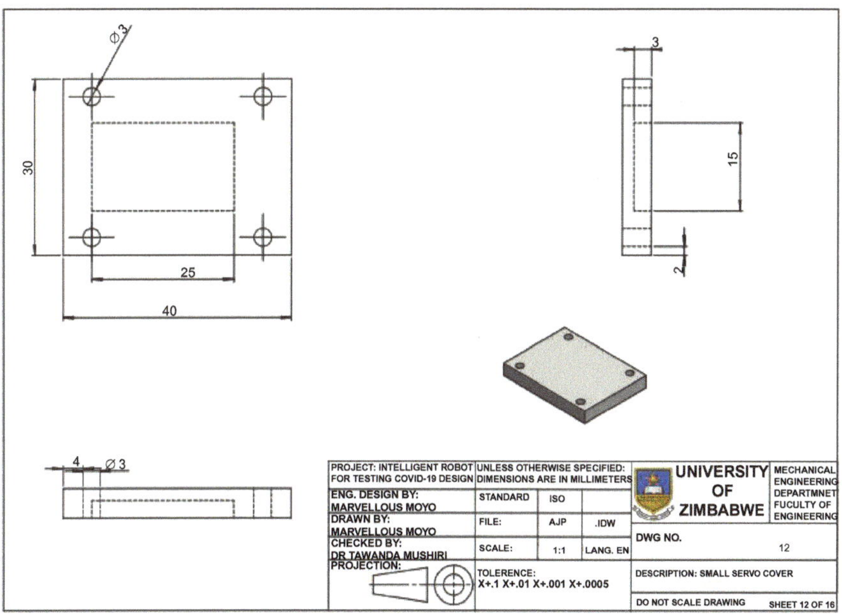
Ø 3
30
25
40
3
15
2
4
Ø 3
PROJECT: INTELLIGENT ROBOT FOR TESTING COVID-19 DESIGN
UNLESS OTHERWISE SPECIFIED: DIMENSIONS ARE IN MILLIMETERS
UNIVERSITY OF ZIMBABWE
MECHANICAL ENGINEERING DEPARTMNET FUCULTY OF ENGINEERING
ENG. DESIGN BY: MARVELLOUS MOYO
STANDARD
ISO
DRAWN BY: MARVELLOUS MOYO
FILE:
AJP
.IDW
DWG NO.
12
CHECKED BY: DR TAWANDA MUSHIRI
SCALE:
1:1
LANG. EN
PROJECTION:
TOLERENCE: X+.1 X+.01 X+.001 X+.0005
DESCRIPTION: SMALL SERVO COVER
DO NOT SCALE DRAWING
SHEET 12 OF 16

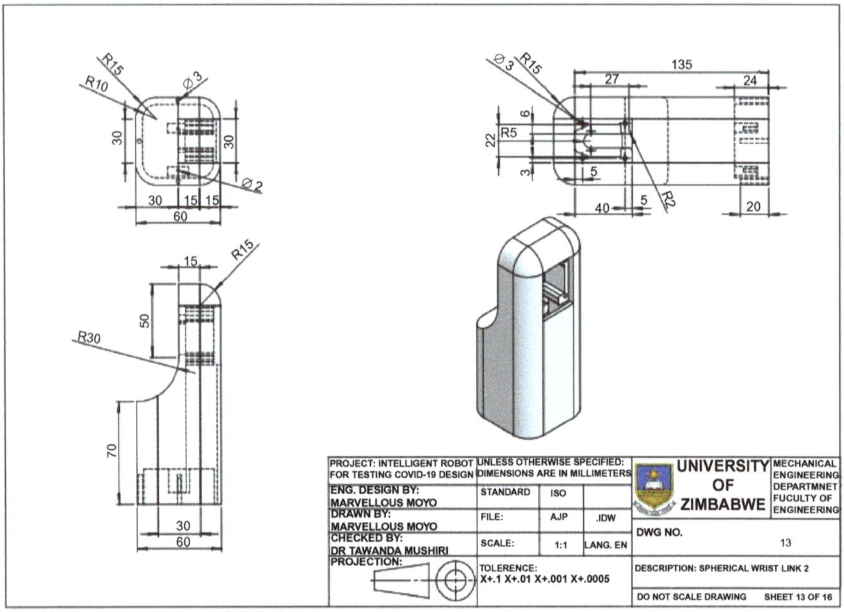
R15
R10
⌀3
30
30
⌀2
30 15 15
60
15
R15
50
R30
70
30
60
⌀3 R15
135
27
24
6
22 R5
3
5
40 5
R2
20
PROJECT: INTELLIGENT ROBOT FOR TESTING COVID-19 DESIGN
UNLESS OTHERWISE SPECIFIED: DIMENSIONS ARE IN MILLIMETERS
UNIVERSITY OF ZIMBABWE
MECHANICAL ENGINEERING DEPARTMNET FUCULTY OF ENGINEERING
ENG. DESIGN BY: MARVELLOUS MOYO
STANDARD ISO
DRAWN BY: MARVELLOUS MOYO
FILE: AJP .IDW
CHECKED BY: DR TAWANDA MUSHIRI
SCALE: 1:1 LANG. EN
DWG NO. 13
PROJECTION:
TOLERENCE: X+.1 X+.01 X+.001 X+.0005
DESCRIPTION: SPHERICAL WRIST LINK 2
DO NOT SCALE DRAWING
SHEET 13 OF 16

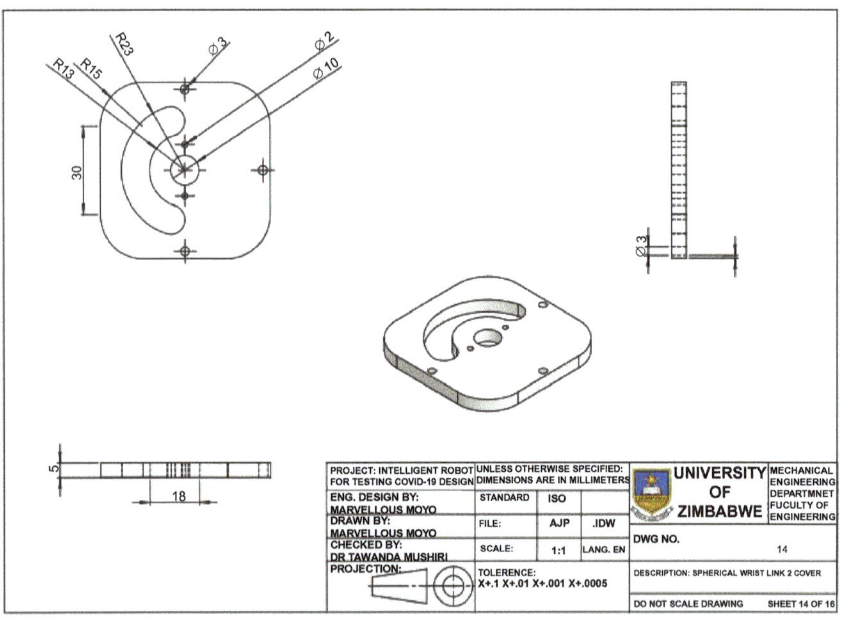
R23
⌀3
⌀2
⌀10
R13
R15
30
⌀3
5
18
PROJECT: INTELLIGENT ROBOT FOR TESTING COVID-19 DESIGN
UNLESS OTHERWISE SPECIFIED: DIMENSIONS ARE IN MILLIMETERS
UNIVERSITY OF ZIMBABWE
MECHANICAL ENGINEERING DEPARTMNET FUCULTY OF ENGINEERING
ENG. DESIGN BY: MARVELLOUS MOYO
STANDARD ISO
DRAWN BY: MARVELLOUS MOYO
FILE: AJP .IDW
CHECKED BY: DR TAWANDA MUSHIRI
SCALE: 1:1 LANG. EN
DWG NO. 14
PROJECTION:
TOLERENCE: X+.1 X+.01 X+.001 X+.0005
DESCRIPTION: SPHERICAL WRIST LINK 2 COVER
DO NOT SCALE DRAWING
SHEET 14 OF 16

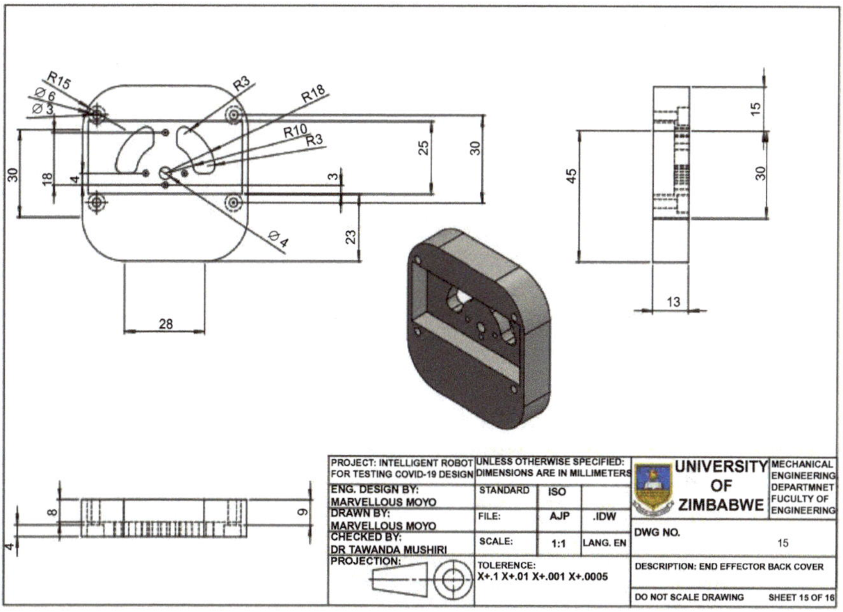
R15
Ø6
Ø3
R3
R18
R10
R3
30
18
4
3
25
30
23
Ø4
28
45
15
30
13
8
9
4
PROJECT: INTELLIGENT ROBOT FOR TESTING COVID-19 DESIGN
UNLESS OTHERWISE SPECIFIED: DIMENSIONS ARE IN MILLIMETERS
UNIVERSITY OF ZIMBABWE
MECHANICAL ENGINEERING DEPARTMNET FUCULTY OF ENGINEERING
ENG. DESIGN BY: MARVELLOUS MOYO
STANDARD ISO
DRAWN BY: MARVELLOUS MOYO
FILE: AJP .IDW
DWG NO. 15
CHECKED BY: DR TAWANDA MUSHIRI
SCALE: 1:1 LANG. EN
PROJECTION:
TOLERENCE: X+.1 X+.01 X+.001 X+.0005
DESCRIPTION: END EFFECTOR BACK COVER
DO NOT SCALE DRAWING
SHEET 15 OF 16

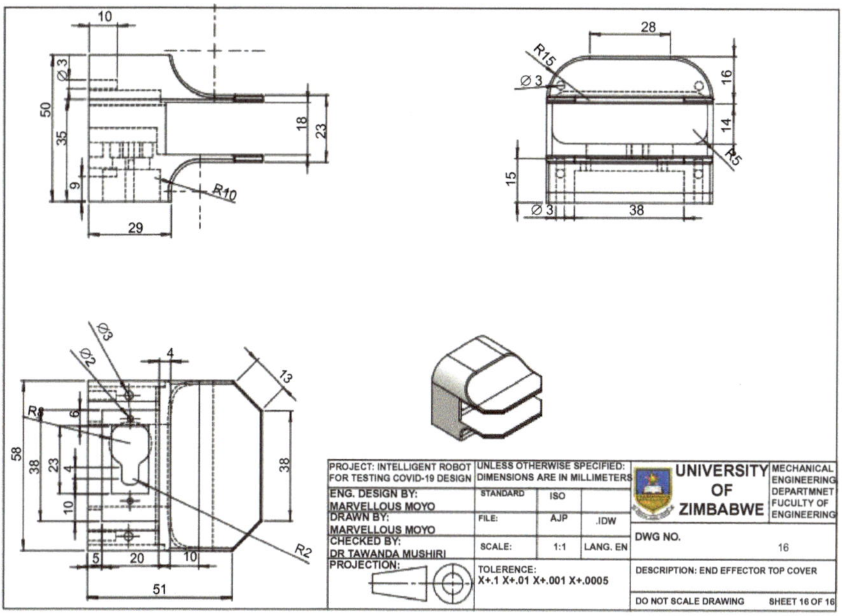
10
Ø3
50
35
9
18
23
R10
29
28
R15
Ø3
16
14
R5
15
Ø3
38
Ø3
Ø2
4
13
R8
6
58
38
23
4
10
38
R2
5
20
10
51
PROJECT: INTELLIGENT ROBOT FOR TESTING COVID-19 DESIGN
UNLESS OTHERWISE SPECIFIED: DIMENSIONS ARE IN MILLIMETERS
UNIVERSITY OF ZIMBABWE
MECHANICAL ENGINEERING DEPARTMNET FUCULTY OF ENGINEERING
ENG. DESIGN BY: MARVELLOUS MOYO
STANDARD ISO
DRAWN BY: MARVELLOUS MOYO
FILE: AJP .IDW
DWG NO. 16
CHECKED BY: DR TAWANDA MUSHIRI
SCALE: 1:1 LANG. EN
PROJECTION:
TOLERENCE: X+.1 X+.01 X+.001 X+.0005
DESCRIPTION: END EFFECTOR TOP COVER
DO NOT SCALE DRAWING
SHEET 16 OF 16

Index

Note: Page numbers followed by *f* indicate figures and *t* indicate tables.

FSC
www.fsc.org
MIX
Papier | Fördert gute Waldnutzung
FSC® C083411

Kolibri 360 GmbH
Ferdinand-Jühlke-Straße 7
99095 Erfurt, Deutschland
produktsicherheit@kolibri360.de